Omar MESSAOUDI

Industrial microbiology course

Omar MESSAOUDI

Industrial microbiology course

ScienciaScripts

Cover image: www.ingimage.com

This book is a translation from the original published under ISBN 978-620-6-71452-1.

Publisher:
Sciencia Scripts
is a trademark of
Dodo Books Indian Ocean Ltd. and OmniScriptum S.R.L publishing group

120 High Road, East Finchley, London, N2 9ED, United Kingdom
Str. Armeneasca 28/1, office 1, Chisinau MD-2012, Republic of Moldova, Europe
Printed at: see last page
ISBN: 978-620-8-08587-2

Foreword

This industrial microbiology book is aimed at students of microbiology, biochemistry and biotechnology. Divided into two parts, it aims to provide fundamental knowledge of industrial microbiology. The first part focuses on the three main players in industrial microbiology: industrial microorganisms, industrial fermenters and industrial culture media. The second part deals with the various products of industrial fermentation, such as microbial biomass, used as a source of proteins of unicellular origin, and the industrial production of primary and secondary metabolites, including amino acids, organic acids, biogas, as well as antibiotics, polysaccharides and vitamins. The pedagogy adopted is based on simple, accessible language, supported by examples, and is enriched by numerous illustrations in the form of simplified diagrams, photos and summary tables to make the content more comprehensible to students.

Table of contents

Introduction .

The use of microorganisms for human well-being is a very old practice, dating back to 7000 BC, with the Sumerians and the Egyptians. During this period, microorganisms were used empirically to preserve foodstuffs, prepare bread, make vinegar and alcoholic beverages, and make cheese **(Baltz et al, 2010).**

Following the discovery of microorganisms, and the knowledge and experience accumulated over time, a new field has emerged, called industrial microbiology. This is a branch of applied microbiology in which microorganisms of interest are exploited for commercial purposes to carry out large-scale biosynthesis, biotransformation or degradation processes. Indeed, industrial microorganisms can be used to produce protein-rich biomass, or for the production of molecules useful to man, derived either from primary metabolism, such as amino acids, organic acids and biogas, or from secondary metabolism, such as antibiotics and polysaccharides. They can also be used to carry out certain biotransformation reactions of toxic organic molecules in order to render them harmless **(Messaoudi et al, 2015; Harwood et al, 2018).**

A large part of the success of using microorganisms in industry is due to their ease of cultivation, on the one hand, and their high growth rate, on the other, which reduces both cost and production time. This overcomes many of the problems encountered when using plants, for example. Indeed, plants used as sources of commercial molecules of interest require a very large cultivation field, more time, as well as suitable climatological conditions (rainfall, temperature, humidity levels, etc.), which increases the cost of production **(Wilson et al 2019).**

Many microorganisms are used in industrial microbiology; these include bacteria, archaea, yeasts, molds and microalgae. These microorganisms can be used either in their natural state, as mutants selected in the laboratory, or as genetically modified organisms (GMOs) **(Zhang et al, 2014).**

I. The fields of activity of industrial microbiology and the benefits of using microorganisms in industry.

1. Introduction.

Fermentation industry is a generic term applied to commercial processes that rely on the ability of microorganisms to produce microbial biomass, useful molecules or catalyze a biotransformation reaction on a large scale, in the presence or absence of oxygen. Thus, contrary to its biochemical meaning, the term industrial fermentation does not refer specifically to the metabolism of the microorganism **(Humphrey et al, 1992).**

Indutrial fermentation takes place in fermenters under rigorously controlled physicochemical conditions, including temperature, pH, aeration, different sources of carbon and nitrogen, etc. **(Walker et al, 2014).**

2. The benefits of using microorganisms in industry.

The use of microorganisms in industry provides a wide range of services. They have proved particularly useful for **(Demain et al. (2008):**

 Easy to grow;
 Their very short generation time ;
 Their high growth rate ;
 Their ability to be grown on inexpensive substrates, often by-products from the food industry;
 Their ease of genetic manipulation.

3. The objectives of industrial fermentation .

The main objectives sought when applying microorganisms in industry are as follows:
 Collect microbial biomass, as in the case of baker's yeast, *Saccharomyces cerevisiae* **(Ritala et al, 2017).**
 Produce metabolites from primary or secondary microbial metabolism, such as antibiotics, amino acids, organic acids **(Keller et al, 2019).**
 Performing biotransformation reactions, consisting in transforming complex or toxic chemicals into simple or less toxic chemical molecules under the action of a microorganism, such as in the case of wastewater treatment **(Huang et al, 2017).**

Other objectives may be pursued depending on the field of application of the microorganisms.

In the food industry, the benefits of fermentation of a food product using industrial microorganisms is to improve their stability, thanks to two mechanisms:
 The consumption of fermentable substrates, usually sugars such as lactose in milk, prevents their post-fermentative use by spoilage bacteria ;
 The production of alcohol or organic acid during fermentation provokes a lower pH, which limits the multiplication of pathogenic bacteria.

Fermentation's vital role in preserving the product is accompanied by positive effects in terms of texture and flavour.

4. Industrial microbiology .

Industrial microbiology has a wide range of applications:

 Agri-food sector: Many food products are the result of industrial fermentation. Yogurt, for example, is made through the action of two bacteria, *Lactobacillus bulgaricus* and *Streptococcus thermophilus*. The lactic acid bacterium *Lactococcus lactis* is used in the manufacture of various types of fresh cheeses, acidifying lactose to form lactic acid, thus facilitating curd formation. Organic acids such as citric acid are also produced industrially by microorganisms such as the fungus *Aspergillus niger*.

 Pharmaceutical field: Microorganisms, in particular *Actinobacteria*, especially the genus *Streptomyces*, are a major source of antibiotics, antifungals, anticancer agents and other pharmaceuticals already on the market. Indeed, around 70% of antibiotics used to treat microbial infections originate from bacteria of the *Streptomyces* genus.

 Bioremediation and environmental remediation field: Microorganisms can be used for the biotransformation of toxic pollutants in decontamination processes known as bioremediation.

 Renewable energy field: Microorganisms play a key role in the field of renewable energy. Indeed, several microorganisms are involved in converting organic waste into biogas (methane) or bioethanol, which are then used as fuels to replace fossil energy.

Other applications are described in **Table 1.**

Table 1. Some examples of the application of microorganisms in industry

Microorganisms	**The product of fermentation**	**The application**
Saccharomyces cerevisiae	Ethanol	Fine chemicals
Pseudomonas	2 Ceto-gluconic acid	An intermediary in the production c ascorbic acid (vitamin C); precursor to the synthesis of vitamin C. isoascorbic.
Aspergillus niger	Pectinase, protease	Fruit juice clarification
Bacillus subtilis	Amylase	Preparation of modified starch ; Use during gluing paper.
Bacillus subtilis	Protease	Protein hydrolysis
Micrococcus glutamicus	Lysine	Food additive
Leuconostoc mesenteroides	Dextran	Food stabilizer
Gluconobacter suboxydans	Sorbose	Ascorbic acid production
Streptomyces olivaceus	Cobalamin (Vitamin B12)	Food supplements
Recombinant *E. coli*	Insulin	Human medicine
Streptococcusthermophilus , *Lactobacillus bulgaricus*	Yogurt	Starter in the dairy industry
Candida utilis *Fusarium* graminearum	Proteins of origin unicellular (P.O.U.)	Human and animal nutrition
Cephalosporium acremonium *Saccharopolyspora erythrea*	*Cephalosporin* *Erythromycin*	Antibiotics for the treatment of infections

II. Industrial microorganisms: isolation, selection and improvement of strains.

1. Introduction.

Microorganisms are widely used in industry for applications useful to mankind. They have proved particularly useful due to their ease of cultivation, rapid growth, ability to use cheap substrates (which in many cases are waste from the food industry), and ability to undergo genetic manipulation easily **(Becker et al., 2017).** The microorganisms used in industry are generally :

 Bacteria: Several bacteria can be used in the industrial field, such as actinomycetes of the *Streptomyces* genus, which alone provide 70% of the antibiotics used **(Al Farraj et al., 2020).** Lactic acid bacteria, such as *Lactobacillus bulgaricus* and *Streptococcus thermophilus*, are used to make yoghurt **(Yamauchi et al., 2019),** while the species *Corynebacterium glutamicum* and *Propionibacterium shermanii* are characterized by their ability to secrete glutamic acid and vitamin B12, respectively (**Zhang et al, 2020).**

 Fungi: The *Saccharomyces cerevisiae* species is the most widely used yeast in the industrial field, mainly for the production of proteins, recombinant proteins and various biopharmaceuticals **(Gervasi et al., 2018).** While *Aspergillus niger* is a mold used for the industrial production of citric acid **(Steiger et al., 2019).**

 Microalgae: These are photosynthetic microorganisms, which can be eukaryotic or prokaryotic, unicellular or multicellular. Cyanobacteria of the Arthrospira genus are used as dietary supplements to combat malnutrition in the form of dietary spirulina (Spirulina) (**De La Jara et al, 2018**).

2. The strategy followed in the search for new industrial microbial strains.

2.1 Isolation of strains.

During this stage, the aim is to isolate microorganisms that are directly relevant to the desired objective. To achieve this, target microorganisms are isolated from their ecological niche. Samples are then inoculated onto an appropriate selective culture medium to isolate the desired germ. The next step is to purify the various colonies obtained to obtain pure cultures **(Becker et al, 2017).**

2.2. Molecular identification of strains.

Isolated strains then undergo molecular identification to determine their taxonomic position. This is achieved by sequencing 16S rDNA, ITS, or other genes, depending on the nature of the microorganism, whether bacterial or fungal.

(Salam et al, 2020).

2.3. Selection of industrial microorganisms.

After isolation and identification, the resulting microorganisms undergo selection according to their biological and technological suitability **(Zhou et al, 2017; Min et al, 2017; Becheur et al, 2024; Paul et al, 2019).** Selection criteria can be divided into two types:

General criteria:

 Ability to produce the molecule of interest or biomass in a short period of time.

 Non-pathogenic, and must not produce undesirable metabolites such as toxins.

□ Ease of genetic manipulation.

 Ability to grow on i n e x p e n s i v e substrates, such as industrial waste food industry.

 No specific requirements for growth factors, including vitamins and amino acids.

 Retain their genetic stability, especially after freezing or lyophilization.

 Able to withstand various technological processes...

 Bacteriophage resistance: Mainly linked to the presence of restriction enzymes in industrial bacterial strains...

Specific criteria :

These criteria are generally linked to the application of the microorganisms. For example, lactic ferments are selected on the basis of their acidifying and flavouring properties, as well as their ability to produce bacteriocins.

3. Improvement of selected strains.

After strain selection according to general and specific criteria, industrial microorganisms are never used in their natural state and therefore undergo genetic improvement aimed primarily at **(Min et al, 2017):**

a) Increase the yield of the desired product concentration or biomass.

b) Reduce the production of undesirable metabolites.

Two methods are used to improve selected strains: random mutagenesis and DNA shuffling. **(Johansen et al, 2017).**

3.1 Random mutagenesis.

Random mutagenesis is a technique used to induce mutations in the genome ofa microorganism at random. These mutations can lead to changes in the characteristics of microorganisms. Random mutagenesis can be achieved using chemical mutagens, such as ethylmethanesulfonate (EMS) or nitrosoguanidine (NTG), which modify DNA bases and cause replication errors, or by exposure to radiation, such as UV or X-rays, which cause breaks in the DNA. Transposons, mobile genetic elements, can also be used to insert mutations into the genome. Finally, incorrect nucleotides can be inoculated during in vitro DNA synthesis by PCR **(Yang et al, 2018).**

For the latter method, multiple mutations can be introduced by performing several PCR cycles on the same gene using a polymerase with a high error rate and no error-correcting exonuclease activity: this is the Error-prone PCR method **(Figure 1).** However, the number of mutations obtained remains low (around 1 to 2 mutations per 1,000 DNA base pairs, at most) **(Shao et al., 2017).**

This mutation rate can nevertheless be improved and modulated by modifying the PCR reaction conditions compared to those of conventional PCR **(Ye et al., 2020):**

- Use of Mn^{2+} instead of Mg^{2+} as cofactor.
- Use of low concentrations of dNTPs.
- Use of dNTP analogues.
- Use of organic co-solvents (e.g. isopropanol) to reduce polymerase fidelity.

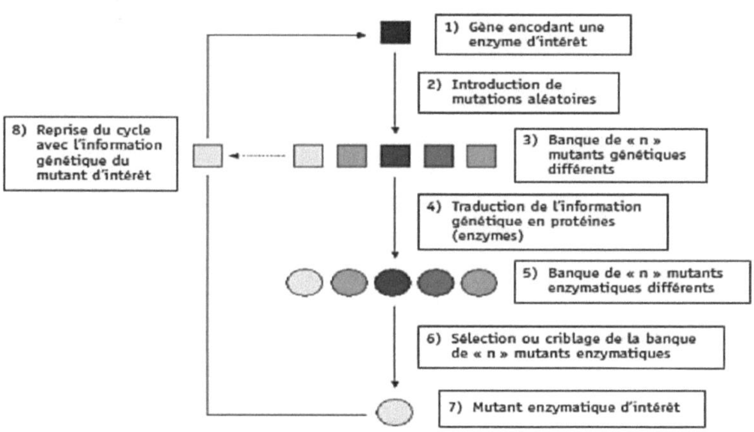

Figure 1. The steps involved in random mutagenesis.

3.2 Breeding genetics.

DNA shuffling refers to the genetic recombination of a series of homologous genes from the same family, or of the same gene with different mutations. The starting point is to mix homologous sequences. The genes are then cut non-specifically and randomly by a DNase I

(Yang et al, 2017). The resulting fragments, 50 to 100 bp in length, will then hybridize to each other according to their sequence homology, then several cycles of primerless PCR are undertaken **(Figure 2).** The result is a hybrid gene library, which will be transformed into host cells in order to search for the desired unique properties of the improved strain **(Brindha et al, 2020).**/

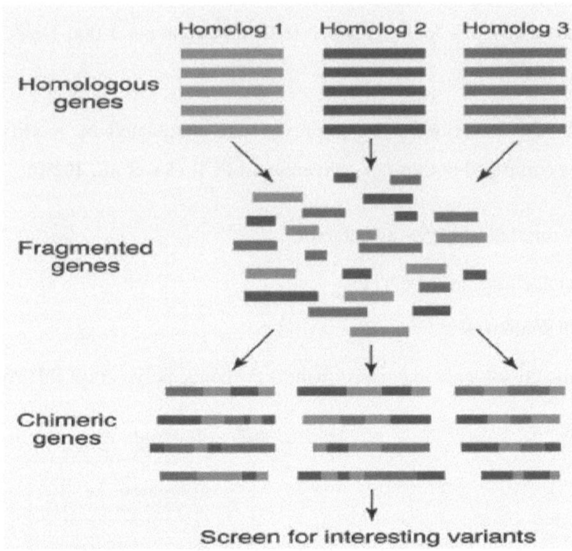

Figure 2. The genetic shuffling method.

3.3 Cloning of mutant or chimeric genes into plasmid vectors .

Most in vitro mutagenesis or recombination processes follow a similar basic pattern: first, the gene of interest undergoes in vitro mutagenesis or recombination, then the resulting copies are introduced into plasmids carrying an antibiotic resistance gene. These plasmids are then introduced by transformation into host organisms (typically bacteria), so that on average an organism receives only one copy. Transformed bacteria are selected by adding an antibiotic to the culture medium, which eliminates all non-transformed bacteria **(Figure 3) (Chang et al., 2017; Asif et al., 2017).**

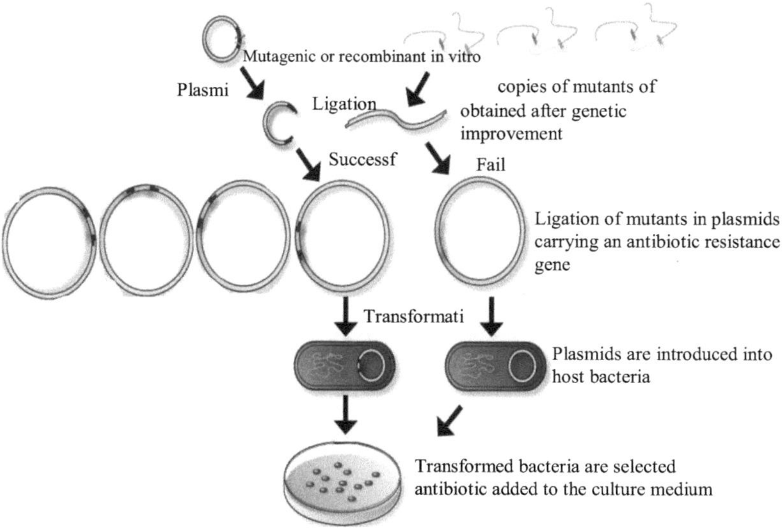

Figure 3. Stages in the genetic improvement of industrial strains **(Chang et al, 2017)**.

Random mutagenesis or genetic recombination results in a heterogeneous population carrying different types of modified genes. Each clone must be tested to identify the modifications of interest. This can be done in two ways:

- If the desired activity is essential to the survival of the transformed cell, it is usually straightforward to identify beneficial modifications by undertaking cell survival selection. For example, if an enzyme is required for the degradation of a specific substrate in the culture medium, only cells with favorable modifications in the gene encoding that enzyme will be able to survive and grow in that medium **(Jang et al, 2013).**
- When selection by cell survival cannot be applied to test the entire mutant library, the experimenter must turn to an alternative technique, screening. Under this approach, all the enzyme mutants in the library must be tested individually to identify those that meet the specific requirements being sought. Conventionally, to undertake a screening, we create a library of 10^2 to 107 mutants in cells, arrange them individually in multi-well plates, and test them one by one for a desired activity or property. In concrete terms, it is possible to follow the evolution of a reagent by staining or by the appearance of fluorescence, to evaluate changes in pH according to an indicator added to the reaction medium, to monitor the turbidity of the solution or even to measure changes in temperature accompanying the reaction of interest to the

experimenter **(Bergmans et al, 1981).**

4. Feasibility .

Pre-selected strains, identified and then improved, need to be propagated with a view to commercialization. Industrial microorganisms need to be produced on a large scale. To achieve this, the strains are first produced at a pilot stage and analyzed at each stage of the manufacturing process (enumeration, viability, etc.) to assess their resistance to the technological process. This makes it possible to determine the fermentation parameters or biomass treatments best suited to the survival of industrial microorganisms. **(Diethard et al, 2017; Kriger et al, 2018).**

Once industrial feasibility has been confirmed, a strain is created. The cells are grown in 2-liter fermenters, then distributed in one-milliliter cryotubes, topped up with a cryoprotectant, and stored under freezing conditions at -80°C, in liquid nitrogen **(Kriger et al, 2018).**

III. Industrial growing environments.

1. Introduction.

Industrial culture media must meet the nutritional requirements of microorganisms to enable optimal growth of cell biomass and, at the same time, provide the nutrients needed for biosynthesis of target metabolites **(Okafor et al, 2020).**

Industrial fermentation is most often carried out on liquid culture media, although some fermentations are carried out on solid culture media **(Demain and 1981).**

2. Factors determining the choice of culture medium industrial.

The main factors affecting the choice of industrial culture media are as follows **(El-Mansi et al, 2018):**

□ Inexpensive substrate.
□ Consistent physico-chemical quality, even after sterilization.
□ Year-round availability.
□ Low transport and storage costs, especially with regard to temperature.
□ Easy to handle in solid or liquid form.
□ Easy to sterilize.
□ Viscosity that does not interfere with agitation or aeration of the industrial culture medium during the fermentation process.

3. The composition of industrial culture media.

The composition of the culture medium must include a source of carbon, which is most often the energy source, as well as a source of nitrogen, phosphorus and sulfur. Trace elements must also be present in the industrial culture medium. Some demanding micro-organisms require the presence of certain growth factors, such as vitamins or amino acids **(Waites et al, 2009).**

3.1. Sources of carbon.

The quantity and quality of the carbon source used as an ingredient in the industrial culture medium can be determined from the biomass yield coefficient (Ycarbon), which is the ratio between the quantity of biomass produced and the quantity of carbon substrate consumed (g biomass / g substrate) **(Waites et al, 2009).**

$$Y_{carbon}(g/g) = \frac{\text{biomass produced (g)}}{\text{carbon substrate utilized (g)}}$$

For reasons of cost, pure carbon sources, such as carbohydrates like glucose or sucrose, are rarely used in industrial-scale fermentations. Instead, other less expensive carbon sources can be used:

 Molasses: This is a by-product resulting from the refining of sugar extracted from sugar beet or sugar cane. It is a dark-colored viscous syrup containing 50-60% (w/v) carbohydrates, mainly sucrose, as well as 2% (w/v) nitrogenous substrates, in addition to some vitamins and mineral salts **(Koval et al, 2019).**

 Malt extract: Aqueous extracts of malted barley can be concentrated to form a syrup very rich in simple sugars and disaccharides, as well as certain vitamins, peptides and amino acids. These substrates can be used to grow filamentous fungi, yeasts and actinomycetes **(Cvetković et al, 2002).**

 Starch and dextrin: Starch is most often obtained from corn, but it can also be obtained from other cereals. To be used as a source of carbon and energy, starch first undergoes hydrolysis to sugar syrup, containing mainly glucose and dextrin, by dilute acids or amylolytic enzymes **(Laluce et al, 1988).**

 Whey: Also known as whey, this is the result of milk coagulation. The composition of whey varies according to the method from which it was obtained. Indeed, whey obtained by rennet is very rich in serum proteins as well as lactose, whereas whey obtained by lactic fermentation of milk is rich in serum proteins and calcium **(Pescuma et al, 2015).**

3.2. Source of nitrogen.

Most industrial microorganisms can utilize nitrogen sources in both organic and inorganic forms. The quantity and quality of the nitrogen source used can be determined from the biomass yield coefficient (Y_{azote}).

Inorganic nitrogen sources can be supplied in the form of ammonium salts, such as ammonium sulfate (SO_4^- NH_4^+) and diammonium phosphate ($(NH_4)_2HPO_4$), while organic nitrogen sources include amino acids, proteins and urea **(Kampen et al, 2014).**

The source of organic nitrogen is often provided in crude form, mainly by-products from the food industry, such as corn steep liquor, yeast extracts, peptones and soybean meal. Pure amino acids are only used in special cases **(Waites et al, 2009)**.

3.3. Salts minerals.

Mineral salts such as cobalt, copper, iron, manganese, molybdenum and zinc are present in sufficient quantities in the water and other ingredients of the industrial culture medium. For example, corn steep liquor contains a satisfactory quantity of mineral salts to meet fermentation requirements **(Zhou et al, 2006).**

3.4. Vitamins and growth factors .

Many industrial microorganisms can synthesize all growth factors, including vitamins, amino acids and fatty acids, from basic nutrients. On the other hand, other demanding, or auxotrophic, microorganisms require the addition of certain types of growth factors to the industrial culture medium **(Chang 1999).**

3.5. O_2 intake .

Depending on the microorganism's type of respiration, oxygen is injected directly into the fermenter in the form of air containing around 21% (v/v) oxygen. On the other hand, in other cases, oxygen can be supplied in pure form when the microorganism's needs are particularly high. In all cases, the oxygen injected into the fermenter must be sterilized by filtration **(Waites et al, 2009)**.

3.6. The addition of antifoams.

Antifoams are needed to reduce foaming during fermentation, which is mainly due to the proteins in the culture medium. If foam formation is not controlled, it can block air filters, resulting in loss of asepsis. There are three approaches to controlling foaming: (i) modifying the composition of the culture medium; (ii) using mechanical defoamers; (iii) adding chemical defoamers, which are surface-active molecules, such as vegetable oils (soybean oil, sunflower oil and rapeseed oil) or fish oil **(Pelton et al, 2002).**

IV. Fermenters or industrial bioreactors.

1. Introduction .

A bioreactor, also called a fermenter or propagator, is a device in which microorganisms are multiplied for the production of biomass, the production of a metabolite, or the bioconversion of a target molecule **(Erickson, 2019).**

A bioreactor comprises **(Figure 4):**

 A glass or stainless steel culture chamber, with a variable volume ranging from a few liters to several cubic meters in the case of industrial units, this tank is hermetically sealed.

 A stirring system used to provide agitation and aeration of the culture, formed by an external motor and one or more internal turbines (depending on fermenter size).

 A syringe for injecting culture medium or nutrients.

 Probes for checking temperature (thermometer), pH (pH meter), and dissolved oxygen concentration (oximeter probe).

 A computer-managed control unit for recording and controlling all operating parameters.

Figure 4: Diagram of an industrial fermenter.

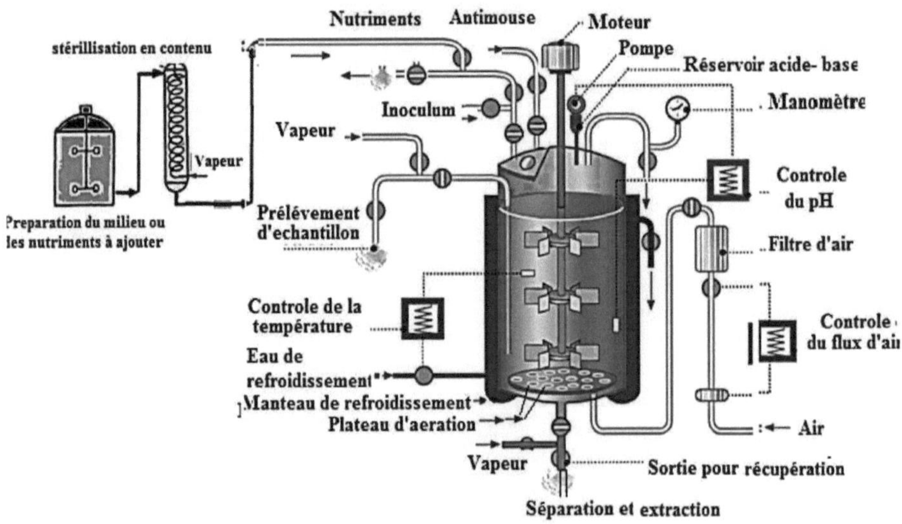

Bioreactors are classified according to their maximum volume:

 18 L volume sterilizable laboratory bioreactors.

 in situ sterilizable laboratory bioreactors up to 30 L ;

 pilot bioreactors up to 300 L;

 industrial bioreactors up to 500,000 L (500 m^3).

2. Types of fermenters.

There are three types of fermenter:

2.1. Batch mode .

In this type of fermenter, the system is closed and maintains a constant volume. The tank is filled with sterile culture medium, then inoculated with the industrial strain. Fermentation takes place under stirring, and throughout the entire fermentation process, the culture volume remains constant without any additional introduction of culture medium **(Figure 5)**. However, neutralizing reagents or defoaming agents can be added **(Carmaux, 2008).**

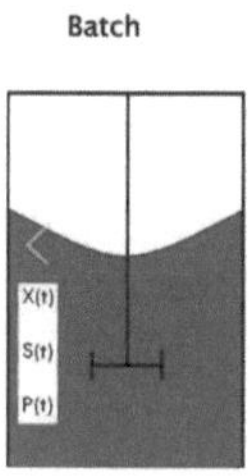

Figure 5: Batch fermenter (**Carmaux, 2008)**.

Biomass concentration increases according to the microbial growth curve **(Figure 6).** At the same time, the substrate is consumed by the microorganism and the concentration of desired products (P) increases. At the end of cultivation, the fermenter is emptied and the desired product extracted **(Oosterhuis et al, 2011).**

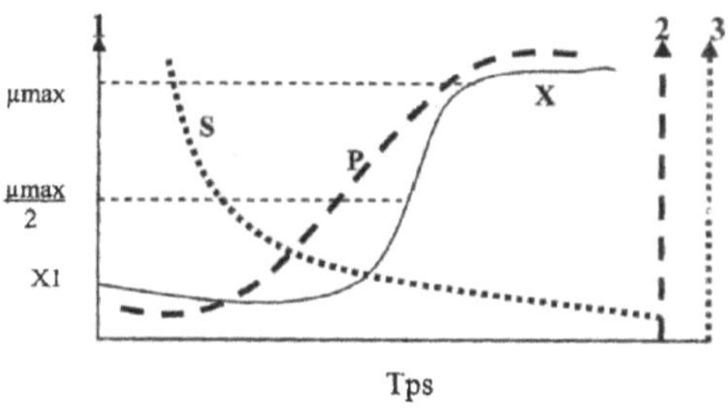

Figure 6. Evolution of microbial cell growth rate (X) vs. substrate utilization (S) and metabolite production (P) during fermentation. **(Oosterhuis et al, 2011).**

The advantages of batch fermentation.

- The desired product can be collected at any time.
- The risk of crop contamination is low.

The disadvantages of batch fermentation.

 The latency time is very long.

 The duration of the exponential phase is very short, so biomass and final product are produced in small quantities, which limits yield.

2.2. Fed Batch (batch fermenter and fed).

In this type of fermenter, in order to shorten the latent phase and prolong the exponential growth phase, cultivation begins with the use of a small volume of culture medium, called the tank foot, inoculated with microbial inoculum. When the microorganism reaches the exponential growth phase, sterile culture medium is introduced into the tank. The feed rate is adjusted to maintain a constant substrate concentration in the tank, without inhibiting biomass production **(Figure 7).** Fermentation is interrupted as soon as the tank is filled with culture medium. It is important to note that the risk of contamination in this type of fermenter is higher **(Evgenios et al, 2020).**

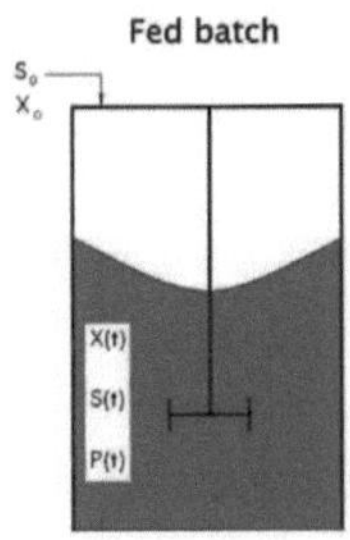

Figure 7. Fed Batch fermenter **(Evgenios et al, 2020).**

2.3. Fermenter in continuous feed mode.

In this type of fermenter, the exponential phase of microbial growth is permanently maintained by regular addition of new culture medium at a constant flow rate. This ensures nutrient replenishment and pH maintenance. At the same time, an equivalent quantity of culture medium must be removed, thus avoiding waste accumulation **(Figure 8) (Sarkar et al, 2003).**

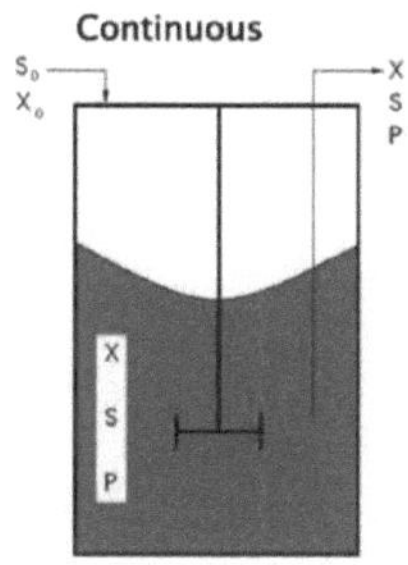

Figure 8. Continuous fermenter **(Evgenios et al, 2020).**

Among the devices used, the turbidostat is a continuous feed fermenter **(Figure 9).** Thanks to turbidimetric control, the concentration of the culture medium is kept constant:

□ If the microbial load tends to increase too much, new medium is added to dilute and bring the microbial cloudiness back to its initial value.

□ If the microbial load decreases, the supply of new medium is reduced until the microbial load is reduced.

to its initial value.

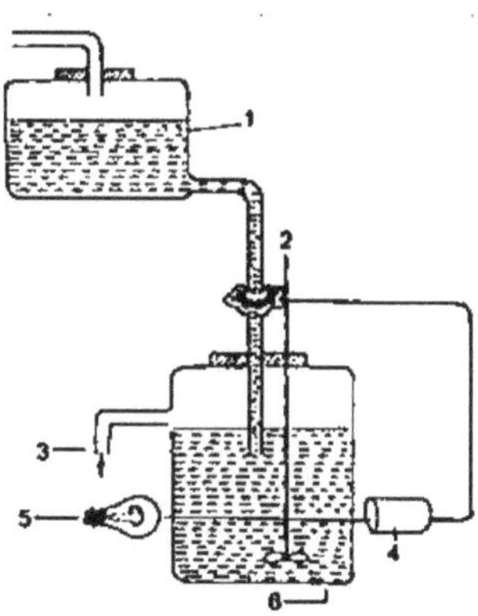

Figure 9. A continuous turbidostat bioreactor **(Lee et al, 2011).**

1: Sterile medium reservoir, **2:** New medium flow control valve, **3:** Culture medium outlet, **4:** Photoelectric cell; **5:** Light source; 6: Fermenter tank.

3. Scaling process (or extrapolation): *scale up.*

Whatever the field of industrial microbiology, the transition from laboratory Erlenmeyer flasks to industrial bioreactors remains a challenge. That's why the process of scaling up is crucial. This involves transferring the microbial culture, prepared in laboratory Erlenmeyers, to small-volume laboratory bioreactors, then to pilot bioreactors, and finally to industrial fermenters (Figure 10). To ensure successful scale-up, the various physicochemical parameters are analyzed and modified during each stage of the scale-up. The physicochemical and enzymatic reactions of microbial cells inside the bioreactor vary according to the reactor volume used. The aim is to achieve the same yield despite the increase in culture volume **(Levin, 2001; Schmidt et al, 2005; Xing et al, 2009).**

Figure 10. Scale-up process **(Levin, 2001).**

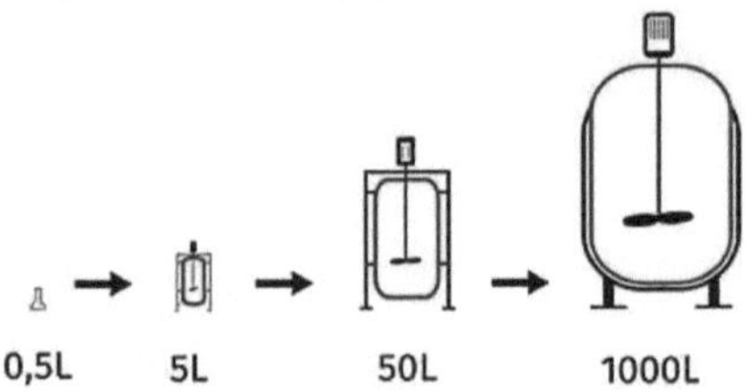

V. The products of industrial fermentation.

V.1. Proteins of Unicellular Origin (P.O.U).

1. Definition.

Any protein-rich microbial biomass destined for human or animal consumption is known as "Protein of Single Cell Origin" (P.O.U.), or Single Cell Proteins (SCP). SCPs are not pure proteins, but also contain carbohydrates, lipids, nucleic acids, mineral salts and vitamins **(Nasseri et al, 2011).**

2. The reason for the world's need for single-cell proteins .

Conventional protein sources are essentially of animal origin, represented mainly by meat of various types. However, because of its high price, the diet of many countries suffers from a serious deficit in animal protein, and the search for new protein resources is one of the concerns of these countries. Thus, the motivation for using P.O.U. in human food is primarily to overcome undernourishment in these countries, thereby meeting their protein needs. In countries where there are no serious problems of undernourishment, the use of P.O.U. for human consumption is very limited and is used more for animal feed. **(Ravindra et al, 2000)**.

3. The microorganisms used to produce P.O.U.

Generally, four types of microorganisms are used: microalgae, bacteria, yeast and filamentous fungi **(Gervasi et al, 2018):**

- *Saccharomyces cerevisiae:* used mainly as a food additive.
- *Candida utilis:* after inactivation by heating, it is used as a nutritious food, rich in protein and free amino acids, and has a light meaty flavor.
- *Fusarium venenatum:* used for the production of Quorn, a mycoprotein-based meat substitute, possessing a taste that resembles chicken meat **(Reihani et al, 2019).**
- Cyanobacteria can be used as dietary supplements, such as spirulina, produced mainly from the species *Arthrospira platensis* or *Arthrospira maxima*, is rich in protein (60% protein) and vitamins, as well as mineral salts and trace elements (Avila-Leon, **et al, 2012).**

4. Microorganism selection criteria for the production of P.O.U.

Before being used as a source of P.O.U, the microorganism must meet certain criteria, in fact, it must be **(Nangul et al, 2021):**

- Non pathogenic
- High protein content.
- High growth rate ;
- Ease of harvesting ;
- Good resistance to variations in production conditions.

5. Industrial production of P.O.U.

- **Growing conditions.**

The ideal ratio for the different sources of carbon, nitrogen and phosphorus used in the culture medium for P.O.U. production is 100/5/1. Incubation temperature is generally between 30 and 35°C, depending on the microorganism, while pH should be maintained between 4.0 and 5.5. A critical parameter is dissolved oxygen concentration. For aerobic fermentations, the medium should be 40% saturated with oxygen **(Waites et al, 2009)**.

- **Fermentation and purification of P.O.U.**

After sterilizing the culture medium, biomass is produced in a fermenter. The P.O.U. is then recovered by multiple centrifugations, then stored in barrels or dried to obtain a powder free of any living cells **(Ghanem, 1992).**

6. Advantages and disadvantages of POU.

The use of microorganisms as a single-cell protein source has certain advantages over protein sources of animal or plant origin **(Nasseri et al, 2011)**:

 Rapid growth rate, compared with livestock.

 High protein content (30-80% of dry weight);

 The ability to use a wide range of cheap (inexpensive) substrates, including organic waste, for their growth.

 Requires little space and water, compared with livestock farming.

 Solves an environmental problem.

 P.O.U. production is independent of climatic variations.

Some disadvantages that may accompany the use of P.O.U.s as a source of proteins **(Fabregas and Herrero, 1985):**

 They can produce toxins or other harmful metabolites.

 The nucleic acid content of POUs limits their use in human nutrition. Indeed, a high consumption of nucleic acids, raises the concentration of uric acid in blood plasma. There is then a risk of urea precipitation in tissues and joints, leading to symptoms analogous to gout disease.

V.2. Metabolites obtained by microbial fermentation .

V.2.1. Amino acids.

1. Introduction.

Many micro-organisms can synthesize amino acids from inorganic nitrogen compounds, via industrial fermentation processes. These amino acids are mainly used as food and feed supplements. However, several amino acids are used as ingredients in pharmaceuticals and cosmetics, and in the chemical industry for the manufacture of polymers, as in the case of glutamic acid, lysine and tryptophan **(Demain et al, 2008).**

In the rest of the course, we'll discuss the example of industrial acid production.
glutamic acid.

2. Example of the industrial manufacture of glutamic acid .

Glutamic acid is a non-essential amino acid, industrially produced in the form of monosodium glutamate (MSG, E621). This molecule is mainly used as a food additive to enhance the taste of many food products, as it is responsible for the umami taste **(Ikeda, 2003).**

2.1. The main microbial strains used.

Glutamic acid-producing bacteria include species that all belong to the *Actinobacteria* phylum, including *Arthrobacter, Brevibacterium, Corynebacterium, Microbacterium and Micrococcus.* They are Gram-positive, immobile, strict aerobic bacteria, auxotrophic for biotin (**Tatsumi, 2012**).

The industrial strains used for glutamic acid production are mutants of the species *Corynebacterium glutamicum, Corynebacterium callunae, Brevibacterium flavum* and *Brevibacterium lactofermentum*. In order to overcome the constraints encountered during the industrial production of glutamic acid, several steps are followed **(Hermann et al, 2003; Ikeda, 2013):**

 Genetic modification of industrial strains aimed primarily at blocking the metabolic pathways that lead to the biosynthesis of undesirable by-products.
 Inhibition of the regulatory mechanism of glutamic acid production by retro-inhibition, in fact, in wild-type strains of *Corynebacterium glutamicum,* when the final product accumulates and reaches a desirable final concentration, the cytoplasmic production of

glutamic acid is inhibited by the retro-inhibition mechanism,

 Decrease the activity of the enzyme complex, a-ketoglutarate dehydrogenase, also called, oxoglutarate dehydrogenase complex (OGDC), which are enzymes of the Krebs cycle, catalyze the conversion reaction of a-ketoglutarate to succinyl-CoA **(Figure 11)**.

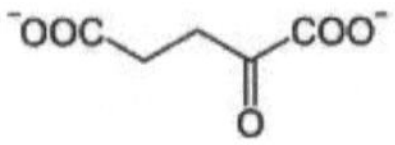

+ CoA-SH + NAD$^+$ → NADH + H$^+$ + CO_2 +

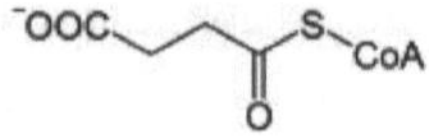

a-KetoglutarateSuccinyl-CoA

Figure 11. Conversion reaction of a-ketoglutarate to succinyl-CoA **(Ikeda, 2013).**

The production of glutamic acid is catalyzed by the enzyme glutamate dehydrogenase **(Figure 12),** which uses oxoglutarate (a-ketoglutarate), a Krebs cycle intermediate, as a substrate **(Ertan, 1992).**

$$\text{Oxoglutaric acid} + NH_4^+ + NAD(P)H + H^+ \xrightarrow{\textit{Glutamatedehydrogenase}} \text{L-glutamic acid} + NAD(P)^+ + H_2O$$

Figure 12. Reaction for the formation of L-glutamic acid **(Ertan, 1992).**

In order to direct the metabolism of mutant strains towards glutamic acid overproduction, the activity of the oxoglutarate dehydrogenase (OGDC) enzyme **(Figure 13)** decreases under biotin-limiting conditions, allowing a-ketoglutarate accumulation and, consequently, leading to improved glutamic acid production yields. It is important to note that the optimum biotin concentration for glutamate production is between 2 and 5 µg/L **(Schultz et al, 2007).**

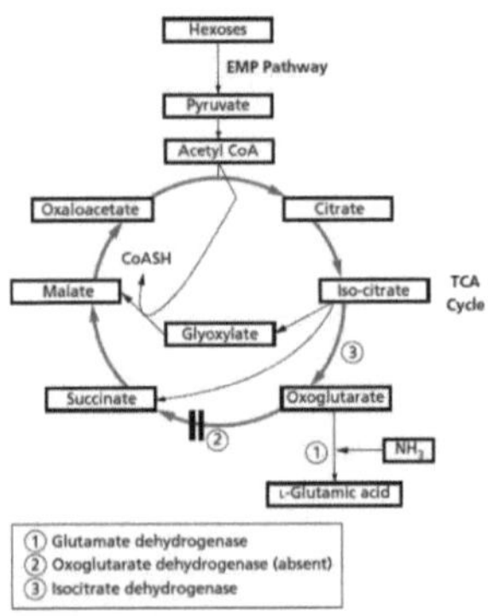

Figure 13. Biosynthesis of l-glutamic acid by mutant strains of *Corynebacterium glutamicum* **(Sonenshein, 2021).**

2.2. The main treatments used to promote acid s e c r e t i o n are glutamic acid.

As glutamic acid biosynthesis is cytoplasmic, and mutant strains do not secrete glutamate, a range of treatments are used to make cells more permeable and facilitate release of the amino acid into the external environment. These treatments include :

 Biotin limitation during glutamic fermentation is responsible for the decrease in membrane phospholipid levels, resulting in an increase in the ratio of saturated to unsaturated fatty acids, which is responsible for glutamate excretion **(Gutmann, 1992).**

 The addition of penicillin appears to be used in the industrial production of glutamic acid. Indeed, penicillin inhibits peptidoglycan biosynthesis in producing strains, which engenders an enlargement of cell size, and at the same time, causes an improvement in glutamate production yield **(Kim et al, 2010).**

 Treatment with surfactants, especially pat Tween 40, in fact, it was observed that final glutamate concentrations of 80 g/L were obtained by the *C. glutamicum* strain, using Tween 40 **(Kim et al, 2009; Hoischen and Kramer, 1990).**

2.3. Growing conditions.

Fermentation takes place in stainless steel fermenters with a maximum capacity of 450 m. The carbon sources used are generally glucose or sucrose. Sugar cane molasses or sugar beet can also be used. The nitrogen source (ammonium salts, urea or ammonia) is fed slowly to avoid inhibition of L-glutamate production **(Hirasawa et al, 2016).**

The culture is maintained under aerobic conditions with agitation, at a temperature of 30-37°C, depending on the microorganism used. To prevent the pH of the culture medium from decreasing as L-glutamate is excreted into the medium, the pH is maintained between 7 and 8. Fermentation normally lasts between 35 and 40 h, reaching broth lglutamic acid levels of 80 g/L **(Delaunay et al, 1999).** Purification of glutamic acid involves centrifugation to remove microbial biomass. The pH of the supernatant is gently lowered by hydrochloric acid to the isoelectric point of L-glutamic acid (pH= 3.2). The L-glutamic acid crystals are then recovered by centrifugation and washed several times. MSG is prepared by adding sodium hydroxide solution to crystalline l-glutamic acid, followed by recrystallization **(Schultz et al, 2007).**

V.2.2. Organic acids .

1. Introduction .

Organic acids are organic molecules with acid groups (COOH) in their carbon skeleton. Most organic acids are used in the food industry, mainly as preservatives, flavorings, acidulants and antioxidants. Among the main organic acids useful to humans are citric acid, acetic acid, lactic acid and propionic acid **(Sauer et al, 2008).**

In the rest of the course, we'll discuss the example of industrial acid production.
citric.

2. Definition and use of citric acid.

Citric acid, or citrate, is an organic acid found in abundance in lemons, hence its name. It is widely used in the food industry, as a food additive (number E330), as an acidulant and flavoring agent in beverages, confectionery, sour candies and other foods. Citric acid can have other non-food applications, especially in medicine and pharmaceuticals **(Max et al, 2010).**

3. Chemical structure of citric acid.

It's a tricarboxylic acid (possessing three carboxyl functional groups: three COOH), plus a hydroxyl (OH) group **(Menzel et al, 2013):**

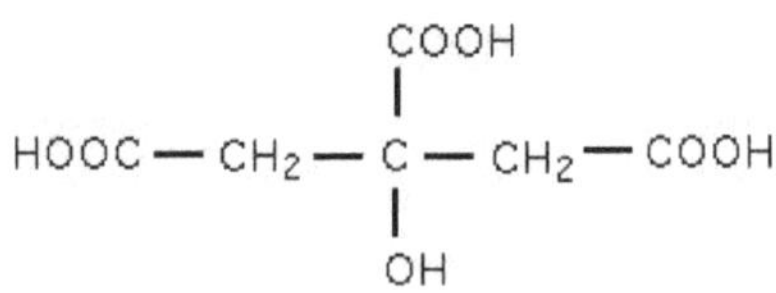

4. Industrial microorganisms producing citric acid.

Many microorganisms can be used to produce citric acid:

- Bacteria: *Bacillus licheniformis* and *Bacillus subtilis* **(Xu et al, 2005).**
- Molds: *Aspergillus niger, Aspergillus awamori, Aspergillus foetidus,* and *Penicillium restrictum* **(Show et al, 2015).**
- Yeasts: *Candida lipolytica, Candida intermedia, and Saccharomyces cerevisiae* **(Yalcin et al, 2010).**

However, worldwide industrial production is ensured by the mold *Aspergillus niger,* due to its ease of handling, and its ability to ferment a variety of inexpensive substrates, as well as their high citric acid production yields **(Papagianni, 2007).**

5. Biosynthesis of citric acid.

Citric acid is a metabolic intermediate of the Krebs cycle, resulting from the condensation of the acetyl residue (2 carbons) of acetyl-CoA onto oxaloacetate (four carbons) to form citrate (six carbons), with the release of CoA. The reaction is catalyzed by the enzyme citrate synthase **(Figure 14) (Szczodrak, 1981):**

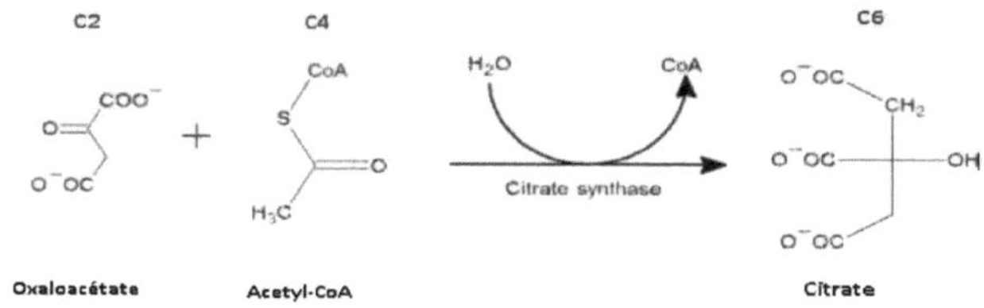

Figure 14. Citric acid formation reaction **(Szczodrak, 1981).**

6. Biochemical aspects of the industrial biosynthesis of citric acid.

Citric acid is a Krebs cycle metabolite intermediate, **and** once formed is immediately converted to cis-aconitate by the enzyme aconitase **(Figure 15) (Degu et al, 2011):**

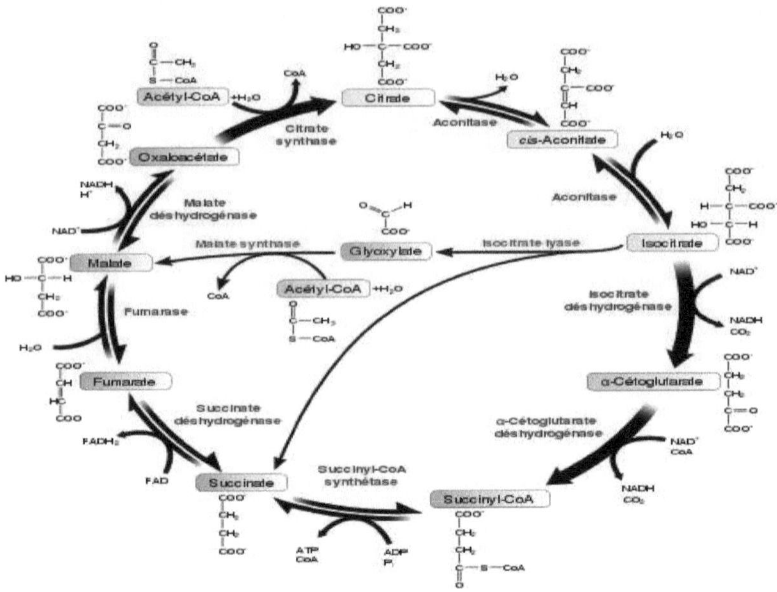

Figure 15. Krebs cycle **(Degu et al, 2011).**

In industry, in order to accumulate citrate, the reaction converting citrate to cis-aconitate must be blocked. This is achieved by inhibiting aconitase, the enzyme that catalyzes this reaction. Inhibition of this enzyme can be achieved in a number of ways, including **(Ramakrishnan et al, 1955; Kubicek and Röhr, 1985):**

- Removal of iron ions (Fe^{+2}), a s the aconitase enzyme uses iron ions as a cofactor, so if iron is removed from the industrial culture medium, the enzyme will automatically be blocked. Iron ions can be eliminated by adding certain iron-chelating agents such as fluorocitrate.
- Genetic modification of *Aspergillus niger* strains, by inducing mutations in the gene responsible for biosynthesis of the aconitase enzyme, in order to obtain transgenic strains lacking this enzyme.

7. The industrial culture medium used for citric acid production.

The process of industrial fermentation o f citric acid involves the cultivation of the mold, *A. niger*, under aerobic conditions, in a culture medium containing :

 A source of carbon.

The carbon source could be starch, starch hydrolysate, sugarcane juice,glucose, sucrose or molasses. In industry, the most commonly used carbon source is molasses. For good citric acid yield, the sugar concentration in the culture medium must be at least 140 g/L (14%) **(Ikram et al, 2004).**

 A source of nitrogen.

These are generally ammonium salts, especially ammonium sulfate, ammonium nitrate, etc., which are generally supplied at concentrations of 0.1 to 0.4 g/L **(Papagianni et al, 2005).**

 Mineral salts.

Mineral salts, especially Fe^{+2} , must be removed from the culture medium, as the latter inhibits citric acid formation above a critical concentration. Mineral salts are removed by ion exchange chromatography **(Choudhary et al., 1966).**

 pH.

To initiate the growth of *Aspergillus niger,* the initial pH of the culture medium is generally between 5 and 7. Thereafter, it must be maintained below 2, which has the advantage of controlling contamination and inhibiting the formation of oxalic acid and gluconic acid (undesirable products) **(Papagianni et al, 2005).**

8. The stages in the industrial production of citric acid.

8.1. Fermentation.

Citric acid can be produced industrially either by surface fermentation or by submerged fermentation:

 The surface fermentation process :

This method involves placing the sterile culture medium, which is usually molasses, in shallow trays (5 to 20 cm deep) made of aluminum or stainless steel, **(figure 16),** placed in a sterile enclosure **(figure 17) (Kılıç et al, 2002).**

The trays are inoculated by spraying them with *Aspergillus niger* spores. The fungus then grows on the surface of the medium. Aerobic conditions are maintained by sterile air blown into the fermentation chamber. The temperature is set at 30°C. The pH gradually drops below 2, at which point citric acid production begins. Fermentation lasts around 8 to 12 days and reaches a yield of around 1.0 kg / m^3 per day **(Darouneh et al, 2009).**

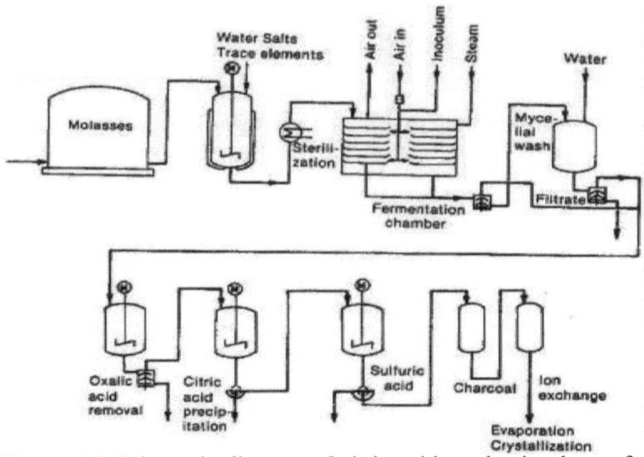

Figure 16. Schematic diagram of citric acid production by surface fermentation **(Darouneh et al, 2009).**

Figure 17. Photos of a citric acid production plant using surface fermentation **(Kılıç et al, 2002).**

 The submerged fermentation process :

Over 80% of the world's citric acid is produced by submerged fermentation in fermenters ranging from 200 to 900 m^3 in volume. Fermenters are corrosion-resistant, generally made of stainless steel. **(Darouneh et al, 2009).** Unlike surface fermentation, the sterile culture medium is seeded with vegetative cells, rather than *A. niger* mold spores. The culture is maintained under agitation at a temperature of 30°C and a pH not exceeding pH=3.5. Fermentation lasts five to fourteen days and achieves a yield of around 18.0 kg/m³ per day **(Khan et al, 1990).**

8.2. Precipitation, extraction and purification of citric acid.

Once fermentation is complete, citric acid purification begins by separating the fungal mycelium from the culture medium by filtration or centrifugation. The resulting solution is heated and mixed with lime (CaO) to form a precipitate of calcium citrate ($Ca_3(C_6H_5O_7)_2$). This is separated by filtration, then treated with dilute sulfuric acid to generate citric acid plus a precipitate of calcium sulfate (gypsum). The latter is removed by filtration. The citric acid solution obtained is decolorized by activated carbon, then evaporated to produce citric acid crystals. These crystals are recovered by centrifugation, then dried and packaged **(Vandenberghe et al, 1999).**

The stages involved in the industrial fermentation of citric acid are summarized in **figure 18.**

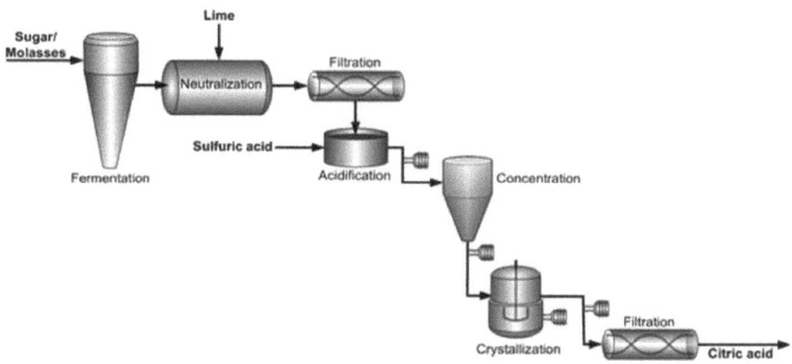

Figure 18. Steps in the manufacture of citric acid.

V.2.3. Biogas production through the methanization process.

1. Introduction.

Methanization (or anaerobic digestion) is the natural biological process of degrading organic matter, in the absence of oxygen (anaerobic conditions), into biogas. Biogas is composed mainly of methane (CH_4) and carbon dioxide (CO_2) **(Lehtomäki et al, 2006).** The anaerobic digestion process takes place in several stages, through the intervention of several microbial communities, in anaerobic digesters or methanizers **(Sagagi et al, 2009).**

2. The benefits of methanization.

Methanization has two main benefits: economic and environmental.
(Martin et al, 2013):

 The economic interest: the methane (biogas) produced can be used as a renewable energy source, which can replace fossil energy sources (especially natural gas), reducing the cost of oil and natural gas. Methane can be used for cooking, or converted into mechanical energy or electricity.
 Environmental benefits: the raw material used in anaerobic digestion generally corresponds to various types of organic waste: urban, agricultural or industrial, which helps to clean up the environment.

3. Biochemical and microbiological aspects of methanization.

Methanization is the result of an extraordinary collective and successive process involving numerous populations of microorganisms. In fact, it's a succession of chemical reactions catalyzed by enzymes produced by various living organisms. Anaerobic digestion (methanization) can be described in terms of four degradation phases **(Figure 19):** hydrolysis, acidogenesis, acetogenesis and methanogenesis. Each phase involves a particular group of microorganisms **(Tian et al, 2017).**

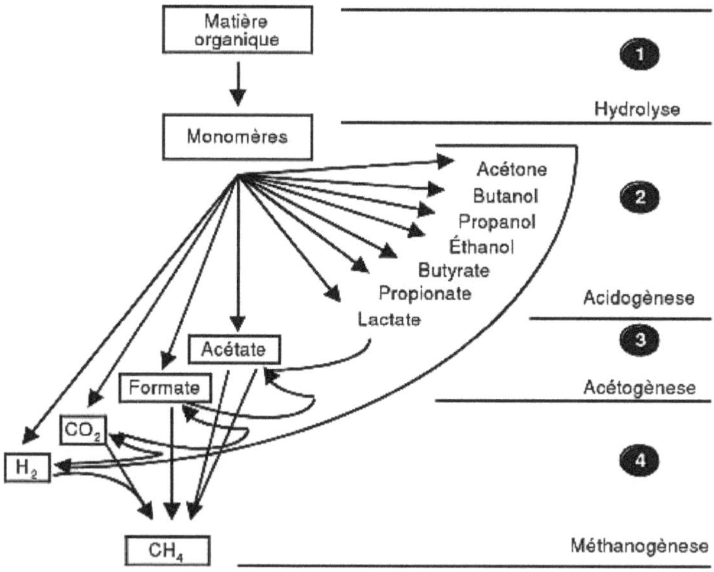

Figure 19. The 4 stages of methanization **(Christy et al, 2014).**

3.1 Hydrolysis.

During this stage, high molecular weight organic molecules such as polysaccharides, lipids, proteins and nucleic acids are hydrolyzed into monomers (monosaccharides such as glucose, fatty acids, amino acids and nitrogenous bases). The bacteria involved in this stage have a strict or facultative anaerobic metabolism and form a heterogeneous phylogenetic group comprising numerous bacterial groups **(Table 2),** including *Clostridium, Bacillus, Bacteroides, Butyrivibrio, Ruminococcus,* Pseudomonas.....etc **(Cirne et al, 2007).**

3.2 Acidogens.

During this stage, the monomers from the hydrolysis step are transformed by fermentation into organic acids (e.g. acetic acid, propionic acid, butyric acid, lactic acid), alcohol (e.g. ethanol), hydrogen and carbon dioxide (**Table 3**). Several bacterial genera are involved in this stage: *Bacteroides*, *Bacillus*, *Pelobacter*, *Acetobacterium*, *Enterobacteriaceae* **(Elefsiniotis et al, 2004).**

Table 2. Hydrolytic bacteria in anaerobic digestion **(Cirne et al, 2007).**

Substrate	Species
Mesophilic conditions	
Cellulose	*Acetivibrio cellulolyticus, Acetivibrio cellulosolvens*
	Clostridium aldrichii, Clostridium cellulolyticum, Clostridium cellobioparum, Clostridium chartatabidum, Clostridium populeti, Clostridium paradoxum, Clostridium cellulofermentans, Clostridium cellulosi Clostridium cellulovorans Clostridium termitidis, Clostridium polysaccharolyticum, Clostridium papyrosolvens, Clostridium josui, Clostridium lentocellum, Clostridium cellulovorans.
	Bacteroides succinogenes , Bacteroides cellulosolvens
	Butyrivibrio fibrisolvens
	Ruminococcus albus, Ruminococcus flavifasiens
	Cilobacterium cellulosolvens
Hemicellulose	*Bacteroides ruminicola*
Pectins	*Clostridium butyricum, Clostridium multifermentens*
Starch	*Bacillus spp.*
	Clostridium butyricum
	Lactobacillus spp.
	Pseudomonas spp.
Lipids	*Anaerovibrio lipolytica*
	Syntrophomonas spp.
Protein	*Bacillus spp.*
	Clostridium spp.
	Staphylococcus spp.
Nitrogen compound	*Clostridium acidiurici, Clostridium cylindrosporum*
	Micrococcus aerogenes, Micrococcus lactilyticus
Thermophilic condition	
Cellulose	*Clostridiumstercorarium , Clostridiumthermocopriae , Clostridium cellulosi, Clostridium thermocellum, Clostridium thermopapyrolyticum*
	Anaerocellum thermophilum
Hemicellulose	*Clostridiumthermocellum , Clostridiumthermocopriae , Clostridium thermobutyricum*
Pectins	*Clostridium thermohydrosulfuricum, Clostridium thermosaccharolyticum, Clostridiumthermocellum , Clostridiumthermolacticum , Clostridium thermosulfurogenes*
	Acetomicrobium faecalis

Starch	*Clostridium thermocopriae Clostridium stercorarium, Clostridium thermolacticum, Clostridium thermosaccharolyticum, Clostridium fervidus, Clostridium thermobutyricum, Clostridium thermopalmarium,*
	Acetomicrobium flavidum
	Thermoanaerobacterfinnii , Thermoanaerobacteracetoethylicus , Thermoanaerobacter ethanolicus, Thermoanaerobacter brockii
Protein	*Coprothermobacter proteolyticus*

Table 3. The different acidogenic reactions **(Elefsiniotis et al, 2004).**

$C6H12O6 + 2\ H2O$	$\rightarrow$ 2 ethanol + $HCO3^-$ + 2 H^+
$C6H12O6$	$\rightarrow$ 2 $lactate^-$ + 2 H^+
$C6H12O6 + 2\ H2O$	$\rightarrow$ $butyrate^-$ + 2 $HCO3^-$ + 3 H^+ + 2 H^+
$C6H12O6$	$\rightarrow$ 3 acetate + 3 H^+
$C6H12O6 + HCO3^-$	$\rightarrow$ $succinate^{2-}$ + $acetate^-$ + $formate^-$ + 3 H^+ + H2O

3.3 Acetogenesis.

In the acetogenesis stage, metabolic intermediates are transformed into acetate, hydrogen and carbon dioxide. These last three will then serve as substrates for methanogens (the next stage in methanization). The acetogenesis stage is catalyzed mainly by homo-acetogenic bacteria, as well as obligatory hydrogen-producing acetogens (which are syntrophic bacteria) **(Francioso et al, 2010).**

3.3.1 Non-syntrophic homo-acetogenic bacteria.

These bacteria produce acetate exclusively. Depending on the origin of the acetate generated, the

homoacetogenic bacteria are divided into two groups **(Kushkevych et al, 2019):**

- **Group 1:** These bacteria produce acetate by fermenting a carbon substrate**.** They can ferment simple oses such as glucose, or products from the acidogenic stage, into acetate. This is the case of: *Butyribacterium* spp. and *Peptococcus glycinophilus.*
- **Group 2:** these bacteria use hydrogen and carbon dioxide to produce acetate :

$4\ H2 + 2\ HCO3^- + H^+ \rightarrow CH3COOH + 4\ H2O$

These include: *Acetoanaerobium noterae, Eubacterium limosum Sporomusa spp. Thermoanaerobacter kivui*

3.3.2. Acetogenic bacteria syntrophs

Acetogenic syntrophic bacteria, or obligate hydrogen producing bacteria (OHPA), are bacteria that produce dihydrogen in the acetate-forming reaction **(Table 4).** Examples include *Syntrophobacter, Syntrophomonas, Syntrophus and Syntrophococcus* **(Schink, 1997).**

The activity of these bacteria is inhibited by an excess of H_2 in the medium, so the hydrogen generated during this stage must be consumed immediately. These bacteria therefore require the presence of H_2-consuming species to ensure that the H_2 pressure is kept low, so the symbiosis of these bacteria with H_2-consuming bacteria, notably sulfate-reducers, homo-acetogenic bacteria and methanogens, is essential to guarantee good activity in the digester **(Huang et al, 2020).**

Table 4. Reactions of syntrophic acetogenic bacteria **(Huang et al, 2020).**

lactate$^-$ + 2 H_2O	→ acetate$^-$ + 2 H_2 + 3 HCO_3^- + H^+
ethanol + 2 HCO_3^-	→ acetate$^-$ + 2 formate$^-$ + H_2O + H^+
ethanol + 2 H_2O	→ acetate$^-$ + 2 H_2 + H^+
butyrate$^-$ + 2 H_2O	→ 2 acetate$^-$ + 2 H_2 + 3 H^+
propionate$^-$ + 3 H_2O	→ acetate$^-$ + 3 H_2 + HCO_3^- + H^+

3.4 The methanogenesis

Methanogens are archaea, using mainly acetate, carbon dioxide and hydrogen as substrates, to generate methane. There are two groups of methanogens **(Lyu et al, 2018; Enzmann et al, 2018):**

- Acetoclastic methanogens: are methanogens that use acetate (acetotrophs) to produce methane. Two archaea genera use this pathway: *Methanosarcina* and *Methanosaeta*.

$CH_3COOH \rightarrow CH_4 + CO_2$

- Hydrogen-trophic methanogens: use hydrogen and carbon dioxide to form methane. Three organisms can use carbon monoxide to generate methane: *Methanothermobacter thermoautotrophicus*, *Methanosarcina barkeri* and *Methanosarcina acetivorans.*

$CO_2 + 4\ H_2 \rightarrow CH_4 + 2\ H_2O$.

Formate can also be used as an electron source by most hydrogenotrophic methanogens:

$4\ HCOOH \rightarrow CH_4 + 3\ CO_2 + 2\ H_2O$.

4. Biogas purification (methane).

For its use as an energy source, biogas undergoes a **(Ferreira et al, 2015):**

- Removal of solid particles and dust,
- Dehydration (to eliminate water).
- Elimination of sulfur compounds (H_2S),
- CO_2 elimination.

V.2.4. antibiotics.

1. Introduction.

Antibiotics are organic molecules derived from the secondary metabolism of filamentous fungi and bacteria, particularly actinomycetes. Although 4,000 antimicrobial molecules have been isolated from various microorganisms, only around 50 are used as antibiotics for the treatment of various infections. Indeed, a bioactive molecule must meet several important criteria to be considered as an antibiotic: (i) : it must be selective, (ii) : it must not be toxic to humans or animals, (iii) : low production costs, (iv) possesses a broad spectrum of activity, (v): a determinate mechanism of action, (vi): high solubility, (x): must be stable vis-à-vis various physicochemical factors ...ect **(Bouzidi et al, 2023; Messaoudi et al, 2020).**

The families of antibiotics most widely used in medicine are b-lactamins, such as penicillins obtained from the mold *Penicillium glaucum*, and cephalosporins purified from the mold *Cephalosporium acremonium*; as well as aminoglycosides, such as streptomycin which was isolated from *Streptomyces griseus* **(Zähner et al, 2020; Messaoudi et al, 2013).**

Since the introduction of penicillin on the world market, bacteria have developed resistance to this organic molecule, and this phenomenon continues unabated. The acquisition of this resistance generally occurs around ten years after the introduction of the antibiotic, and is due to random genetic mutations, or to exchanges of resistance genes between bacteria (genetic transformation, transduction). Antibiotic resistance has made the search for new antibiotic molecules a global emergency **(Messaoudi et al, 2022).**

2. The B-lactam class.

This class of molecules contains antibiotics that contain a β-lactam nucleus formed by three carbon atoms and one nitrogen in its molecular structure **(Figure 20) (El Khoury, 2019):**

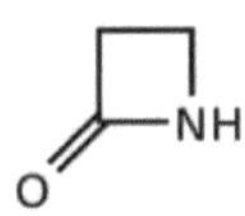

Figure 20. B-lactam core structure **(Suarez, C., & Gudiol, 2009).**

Over 100 B-lactams, mainly penicillins and cephalosporins, have been approved for human use, and they account for more than half of the antibiotics produced worldwide. This group is particularly useful due to its broad spectrum of activity, against a variety of Gram-positive and Gram-positive bacteria **(Page, 2012).**

Among the antibiotics belonging to this family is penicillin. Its structure is characterized by a 4-atom beta-lactam ring fused with a 5-atom heterocyclic ring (thiazolidine). Two methyl groups and a carboxyl group are bonded to this 5-atom ring. The B-lactam nucleus carries an amine function to which various side chains are linked by an amide bond **(Figure 21) (Messaoudi et al, 2021):**

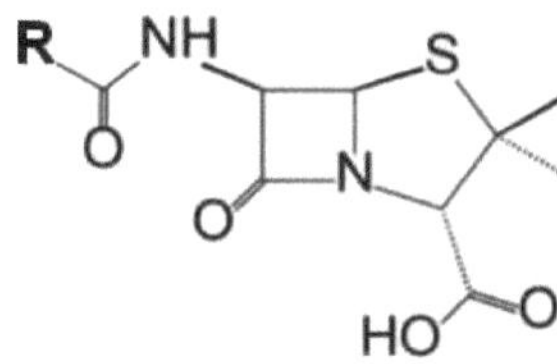

Figure 21. Basic structure of penicillin.

Penicillins act by preventing bacterial wall biosynthesis, following inhibition of the transpeptidase enzyme, a key enzyme in bacterial wall peptidoglycan biosynthesis, and thus cause bacterial death **(Messaoudi, 2020).**

The majority of penicillins are now semi-synthetic, following chemical modification of the natural penicillin obtained from the *P. chrysogenum* species. This modification is achieved by substituting the natural acyl group attached to 6-aminopenicillanic acid (6-APA) with other groups, giving the molecule new properties **(Figure 22) (Parmar et al, 2000).**

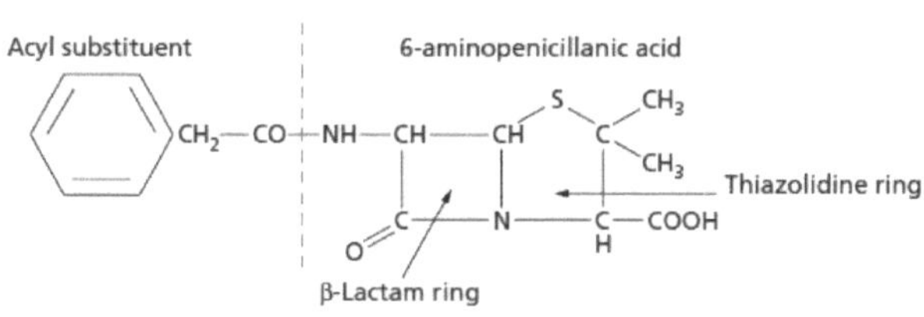

Figure 22. Acyl group substitution of 6-aminopenicillanic acid (6-APA) by a new radical **(Parmar et al, 2000).**

Semi-synthetic penicillins, such as methicillin, carbenicillin and ampicillin, offer various improvements over the original molecule, including resistance to gastric acidity, enabling oral administration, resistance to penicillinase and activity against certain Gram-negative bacteria **(Messaoudi et al, 2019).**

3. Industrial production of penicillin.

The process of obtaining penicillin from the *P. chrysogenum* strain *is* summarized in **Figure 23**. The culture medium used for industrial production of penicillin is formed by **(Veiter et al. 2020; Cepeda-García et al, 2014; Messaoudi et al,2022):**

 A carbon source: which is usually glucose, lactose, sucrose, ethanol and vegetable oils. These carbon sources are used either alone or in combination, as in the case of the glucose/lactose combination. Once glucose has been used up, the *P. chrysogenum* species uses lactose to initiate penicillin biosynthesis.

 A source of nitrogen: the most widely used source of nitrogen is Corn steep liquor.

 Other ingredients may also be added, such a s ammonia, mineral salts, etc.
can also be added.

After preparing and sterilizing the culture medium, this is introduced into Fed-Batch fermenters with a capacity of (20-40) 10^3 liters. The oxygen level should be maintained between 25-60 mmol / L / h, and the culture incubated between 25-27 ° C with a pH of 6.5-7.7 **(Kalai et al. 2014; Abbaszadeh et al. 2014).**

After 6 to 8 days of incubation, the culture is centrifuged to separate the supernatant from the biomass, which will be used for animal feed after several washes to remove traces of penicillin. The penicillin is extracted from the supernatant using the solvent pentyl acetate, from which the penicillin molecule is purified by activated carbon. The antibiotic is then crystallized by the addition of sodium acetate, followed by filtration, and finally, several ethanol washes are performed. Pure penicillin is then dried, and new radicals can be added to its basic structure **(Yadav et al. 2018; Dar et al. 2015; Devi et al, 2009).**

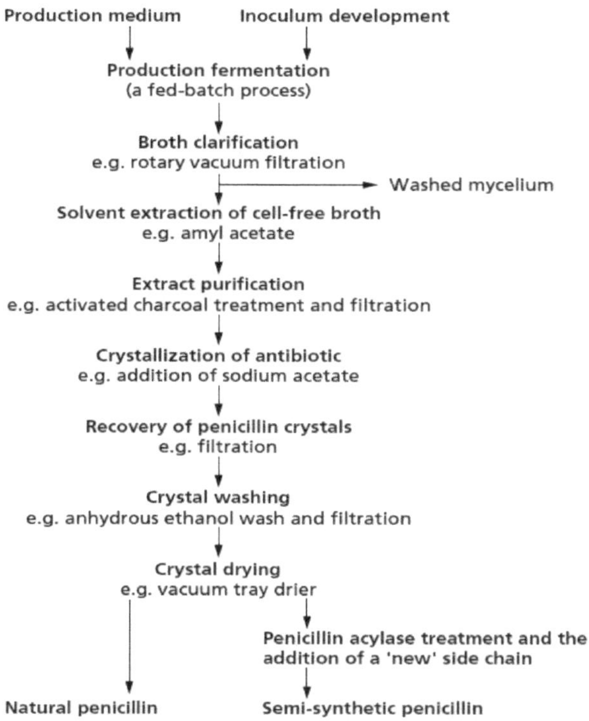

Figure 23. Stages in the industrial manufacture of penicillin (**Devi et al, 2009**).

V.2.5. Polysaccharides of microbial origin

1. Introduction.

Polysaccharides are polymers of single sugars (monosaccharides), linked together by glycosidic bonds. They are formed by a single type of monosaccharide (homopolysaccharide), or are formed by several types of monosaccharides (heterogeneous heteropolysaccharide) **(Salehizadeh et al, 2018).**

Industrial polysaccharides can come from a variety of sources, including plants (cellulose, pectin, gum arabic and starch) and algae (agar-agar, alginate, carrageenan). Microorganisms, including bacteria and fungi, in turn, can synthesize industrial polysaccharides. Depending on their location relative to the producing cell, three types of polysaccharides of microbial origin can be distinguished **(He et al, 2010; Nunez et al, 2013):**

- Intracellular polysaccharides: these types of polysaccharides are difficult to extract;

 Wall polysaccharides: like chitin, which exists in the walls of fungi, these types of polysaccharides are also difficult to extract.

 Extracellular polysaccharides: these types of extracellular polysaccharides do not form covalent bonds with the microbial cell wall. Generally referred to as exopolysaccharides (EPS), they are either secreted out of the microbial cell in the form of a loose slurry, or they encapsulate the microbial cell in the form of a capsule called a slime or "capsular polysaccharides (CPS)".

Exo-polysaccharides of microbial origin fulfill several physiological functions for the microorganism **(Porter et al, 2017; Dumitriu, 2004):**

 They can protect the micro-organism against desiccation,

 Allows the microorganism to escape the immune system,

 act as a barrier against viruses and chemical agents,

 facilitates the attachment of the microorganism to the various surfaces,

 Are energy reserves.

2. Microbial exopolysaccharides and their technological applications.

2.1. Alginate.

It is a linear heteropolymer of D-mannuronic acid, some units of which are acetylated, and L-guluronic acid. Bonding occurs via β-1-4 **(Figure 24).** This polymer is biosynthesized by Gram-negative bacteria, *Pseudomonas* sp and *Azotobacter vinlandii* **(Franklin et al, 2011).**

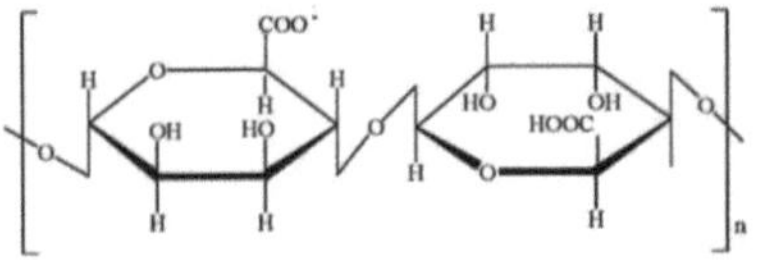

Figure 24. Chemical structure of alginate **(Franklin et al, 2011).**

Alginates are used in the food industry for their emulsifying, gelling, thickening and stabilizing properties, especially in milk-based products (ice creams) and pastries. It can also be used to encapsulate medicines, in the form of alginate beads, and in the manufacture of beauty products **(Bregni et al, 2000).**

2.2. The curdlane

It is an unbranched linear homopolymer of D-glucose, united by β (1→3) glycosidic linkages **(Figure 25).** Produced by the species *Alcaligenes faecalis var.myxogenes*, as well as certain species of the genus *Agrobacterium*. It is used as a food additive (thickener and stabilizer)

(E424) **(Jiang et al, 2013).**

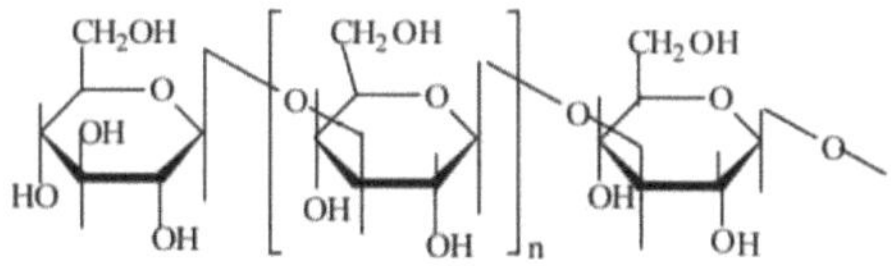

Figure 25. Chemical structure of curdlan **(Jiang et al, 2013).**

2.3. dextran.

Is a branched homopolymer of a-D-glucose, linked by a (1→6) glycosidic bonds, with a (1→3) or a (1→4) branching **(Figure 26).** This polysaccharide is synthesized by several microorganisms, including *Leuconostoc mesenteroides* **(Naessens et al, 2005).**

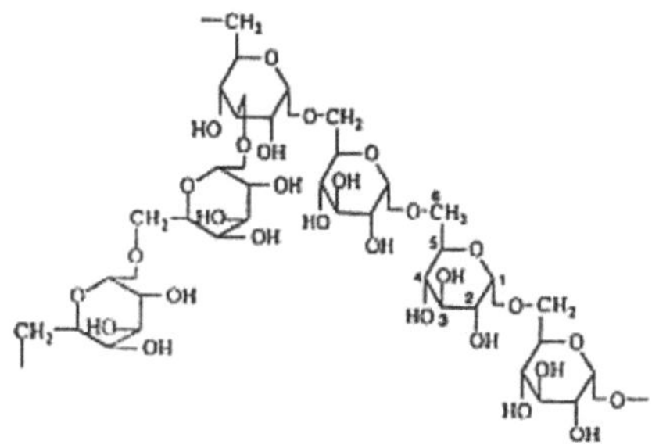

Figure 26. Chemical structure of dextran **(Purama et al, 2009).**

Dextran is produced industrially from sucrose by the enzyme dextranase-sucrase. The resulting product, known as "native dextran", has a high average molar mass, so it cannot be used in this state. For this reason, "native dextran" undergoes mild acid hydrolysis with dilute hydrochloric acid, to obtain a dextran with a low average molar mass, more suitable for the various applications targeted, especially in vivo applications. Indeed, dextran can be used as a substitute for blood plasma, since it has anti-thrombic effects and activates blood flow **(Robyt et al, 2008).**

2.4. Gellan gum (E418).

It's a linear heteropolyoside, made up of repeating tetrasaccharide units (four sugars), consisting of two D-glucose residues plus L-rhamnose and D-glucuronic acid residues **(Figure 27).** Gellan gum is produced by the bacterium *Sphingomonas elodea*. It is used as a food additive, gelling agent and thickener in a number of food preparations **(Nampoothiri et al, 2003).**

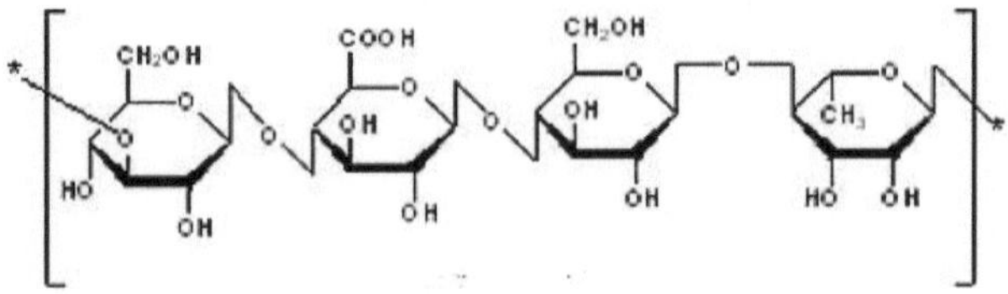

Figure 27. chemical structure of gellan gum **(Nampoothiri et al, 2003).**

2.5. pullulane.

It is a linear heteropolyoside formed by the union of the units maltotriose, which is a triholoside of three glucose molecules connected to each other by a-type (1→4) osidic bonds. The maltotrioses are connected to each other by a-type (1→6) osidic bonds **(Figure 28)** **(Singh et al, 2015).**

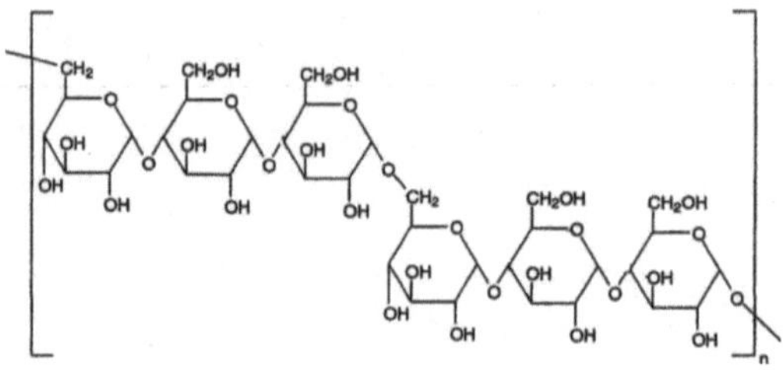

Figure 28. Chemical structure of pullulan **(Singh et al, 2015).**
Pullulan is produced from starch by the yeast *Aureobasidium pullulans*, the main commercial use of pullulan is the manufacture of edible films **(Hamidi et al, 2019).**

2.6. Xanthan gum .

Xanthan gum is a heteropolysaccharide obtained by aerobic fermentation of sugars by bacteria of the genus *Xanthomonas,* including *Xanthomonas campestris*, *Xanthomonas* carotae, *Xanthomonas malvacearum* and *Xanthomonas phaseoli*. Many of these are plant pathogens **(Rosalam et al., 2006).**

These bacteria are small, motile, Gram-negative, strict aerobic bacilli. The *X. campestris* species is used for the commercial production of xanthan **(Leela et al., 2000).**

The structure of xanthan gum is complex, and is formed by a main chain of β-linked D-glucose units (1→4), on which, there are side branches every two glucose units, consisting of a trisaccharide formed by an a-D-mannose, a β-D-glucuronic acid and a terminal β-D-mannose **(Figure 29).** 30% to 40% of β-D-mannoses, from commercially available xanthans, carry a chelated pyruvate moiety between carbons 4 and 6, while 60% to 70% of a-D-mannose is acetylated on carbon 6 **(Sworn, 2021).**

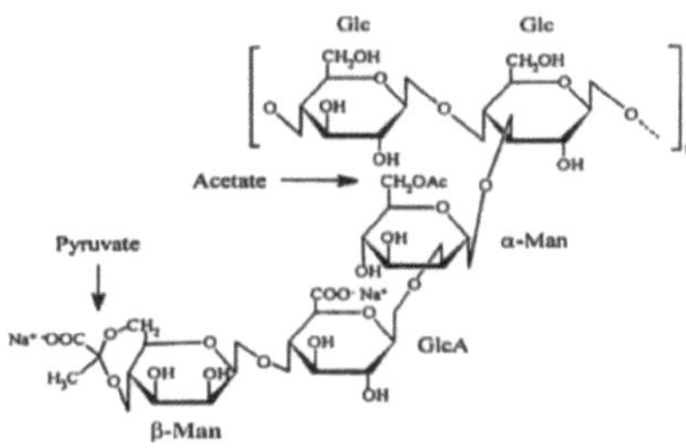

Figure 29. Chemical structure of xanthan gum **(Sworn, 2021)**.

3. Example of industrial polysaccharide production: xanthan gum.

Around 20,000 tonnes of xanthan are produced each year. Production is influenced by several factors, such as the type of fermenter used and its mode of operation, the composition of the culture medium and the physicochemical conditions of fermentation **(Palaniraj et al, 2011).**

The first stage consists of culturing the bacterium, *X. campestris*. The culture obtained will be used to inoculate pilot-scale fermenters of 50 to 200 m^3 volume. The industrial culture medium used for xanthan gum production has the following composition **(Garcıa-Ochoa et al, 2000; Palaniraj et al, 2011):**

- A carbon source: which is added at a rate of 30-40 g / L, it is usually D- glucose, sucrose or molasses.
- A nitrogen source: added with a ratio, carbon/nitrogen of about 10/1, to obtain a high product yield. The main nitrogen sources used are: casein, soy hydrolysate, ammonium salts, peptone, Corn steep liquor, and yeast extract.
- Traces of $MgCl_2$.
- A phosphate buffer solution.

The culture is kept agitated at 28-30°C, at pH=7.0. The bacteria begin to produce xanthan during the exponential phase of growth, and fermentation is normally completed after 3 days of incubation. At the end of fermentation, the culture is heated to 100-110°C for 10 min to kill the bacteria and improve the rheological properties of the xanthan. This is followed by a series of purification steps, including filtration, isopropanol precipitation and centrifugation to obtain xanthan gum. After drying, the residue is ground to obtain xanthan gum powder.

The expected minimum yield is around 25g/L at 50 g/L **(Silva et al, 2009; Gumus et al, 2010).**

V.2.6. The vitamins.

1. Introduction.

Vitamins are organic molecules essential, in minute quantities, to the cellular metabolism of a living organism, which is incapable of synthesizing them. These organic molecules fulfill several physiological roles for the cell: some act as antioxidant molecules, others are enzymatic cofactors, while some, like vitamin D, have a hormonal role. There are two types of vitamins: lipophilic vitamins, such as vitamins D, E, K and A, and water-soluble vitamins, including the eight group B vitamins (for beriberi, a disease caused by vitamin B1 deficiency) **(Chawla and Kvarnberg, 2014).**

Among the water-soluble vitamins is vitamin B12, which is used as a dietary supplement, since it is essential for normal brain function and blood formation, for which reason this molecule is used in the treatment of pernicious anemia, as well as being essential for synthesis, DNA regulation, and amino acid and fatty acid metabolism **(Dziegielewska-Gesiak et al, 2019).**

Several vitamins are currently produced industrially by microbial means, including cobalamin (vitamin B12). This chapter discusses the structure, producing microorganisms and industrial production of vitamin B12 **(Piao et al, 2004).**

2. Structure and microorganisms that produce vitamin B12.

Only bacteria and archaea can synthesize vitamin B12. Bacteria capable of synthesizing cobalamins (vitamin B12), are generally isolated from soil, and belong to different species, including, *Propionibacterium freudenreichii*; *Pseudomonas denitrificans*; *Propionibacterium shermanii*, *Streptomyces olivaceus* **(Piwowarek et al, 2018).**

Structurally, vitamin B12 is an organic molecule composed of a tetrapyrrolic corrinoid ring, with a cobalt atom in the center, which binds with upper (R) and lower (R') axial ligands **(Kräutler, 2005).**

Depending on the nature of the axial ligands (upper and lower), vitamin B12 will take on different names: "cyanopseudocobalamin", "methylpseudocobalamin", "hydroxypseudocobalamin" or "adenosylpseudocobalamin". The lower axial ligand (R') may be a dimethylbenzimidazole (DMB) or adenine group, while the upper axial ligand (R) may be either a cyano (-CN), hydroxyl (-OH) or methyl (-CH3) group **(Figure 31) (Krishnakumar et al, 2007).**

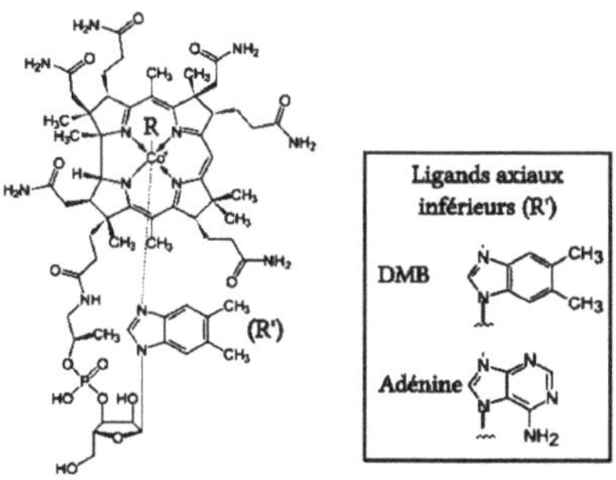

Figure 30. Structure of vitamin B12 **(Hazra et al, 2015).**

3. Industrial production of vitamin B12.

Due to their high vitamin B12 productivity and rapid growth, the species, *Propionibacterium shermanii, Propionibacterium freudenreichii* and *Pseudomonas denitrificans,* are the most widely used in industry for vitamin B12 production. However, due to the different mechanism of vitamin B12 biosynthesis, the fermentation conditions used are different for the two genera **(Hazra et al, 2015).**

Both species, *Propionibacterium shermanii, Propionibacterium freudenreichii,* are microaerophiles, and are able to produce vitamin B12 only at low oxygen pressures. However, biosynthesis of the lower axial ligand (R') dimethylbenzimidazole (DMBI) is carried out under aerobic conditions. Consequently, two steps are required. During the first 3 days of ferm entation, bacteria are grown under anaerobic conditions. This leads to the accumulation of the intermediate, cobinamide, which is the vitamin B12 precursor lacking the DMBI fragment. In the second phase, the culture is lightly aerated for 1-3 days, enabling the bacteria to begin oxygen synthesis of DMBI, and to bind it to cobinamide, to form vitamin B12 **(Yongsmith et al, 1982).**

In parallel to the intracellular production of vitamin B12, both species of the genus *Propionibacterium,* excrete propionic acid and acetic acid which causes a drop in pH, so it's important to neutralize the accumulated acids in order to maintain the culture at pH 7 **(Piwowarek et al, 2018).**

At the end of fermentation, the culture is heated to 80-120°C for 10-30 min at pH 6.5-8.5 to extract vitamin B12. It is then treated with cyanide or thiocyanate, to obtain cyanocobalamin, which is the commercially most common form of vitamin B12, as it is the most stable molecule in the open air and the easiest to crystallize, and therefore to purify once produced by bacterial fermentation **(Hazra et al, 2015)**.

Vitamin B12 is precipitated by adding tannic acid, then purified with adsorption or ion exchange columns **(Piwowarek et al, 2018).**

Glossary.

Auxotrophic bacteria: require the presence of a growth factor for their development. The opposite is called prototrophic.

Bacteriocins: are peptides of ribosomal origin, synthesized by certain bacteria against other bacteria close to the producing strain.

Bacteriophages: viruses that infect only bacteria.

Biomass is a term that signifies the total quantity (mass) of living organisms (plants, animals, fungi, microorganisms) present in a given biotope or location at a given time.

Chelators: are molecules capable of forming a soluble, non-toxic compound (a chelate) with a positive metal ion (Fe^{+2} , Cu^{+2} ....).

Cofactor: non-protein chemical compound required for the biological activity of a protein (enzyme).

Dietary supplement: concentrated source of nutrients (vitamins, minerals, fiber, etc.) designed to compensate for deficiencies in a person's regular diet.

Cryoprotectant: a substance used to protect biological tissues from the very low temperatures used to preserve them.

DNase : enzyme clearing deoxyrobonucleic acids into nucleotides or

Renewable energy: energy sources whose natural renewal is fast enough to be considered inexhaustible on a human timescale.

Restriction enzymes: enzymes capable of cutting a DNA fragment at a characteristic nucleotide sequence called the restriction site.

Umami taste: one of the five basic flavors, along with sweet, sour, bitter and salty.

Archaea: unicellular prokaryotic microorganisms, placed in a domain apart from eukaryotes and bacteria.

Lactic acid bacteria: are Gram-positive, microaerophilic bacteria capable of fermenting lactose into lactic acid.

Lyophilization: this is the drying of a product by the direct passage of water from the solid state (frozen) to the vapor state (sublimation).

Surfactant molecules: molecules capable of reducing the surface tension between two surfaces.

Latency phase: time needed for the bacteria to synthesize enzymes adapted to the new polynucleotide substrate.

Plasmids: extracellular DNA molecules

Emulsifier: molecule used to stabilize a mixture of two chromosomal substances, capable of autonomous replication.

non miscible liquids (emulsion).

isoelectric point: or pHi, this is the pH at which the electric charge

Genetic transformation: integration of a foreign DNA fragment into a cell, resulting in phenotypic changes in the recipient organism.

Fossil energy: energy produced by burning coal, oil or natural gas, derived from the transformation of organic matter buried in the ground for millions of years. the molecule's net weight is zero.

Transduction: transfer of genetic material between bacteria, under the action of a bacteriophage.

Umami taste: one of the five basic flavors, along with sweet, sour, bitter and salty.

References.

1. Abbaszadeh, S., Sharifzadeh, A., Shokri, H., Khosravi, A. R., Abbaszadeh, A. (2014). Antifungal efficacy of thymol, carvacrol, eugenol and menthol as alternative agents to control the growth of food-relevant fungi. Journal de Mycologie Medicale, 24(2), e51-e56.

2. Al Farraj, D. A., Varghese, R., Vágvölgyi, C., Elshikh, M. S., Alokda, A. M., & Mahmoud, A. H. (2020). Antibiotics production in optimized culture condition using low cost substrates from Streptomyces sp. AS4 isolated from mangrove soil sediment. Journal of King Saud University- Science, 32(2), 1528-1535.

3. Asif, A., Mohsin, H., Tanvir, R., & Rehman, Y. (2017). Revisiting the mechanisms involved in calcium chloride induced bacterial transformation. Frontiers in microbiology, 8, 2169.

4. Avila-Leon, I., Chuei Matsudo, M., Sato, S., & De Carvalho, J. C. M. (2012). Arthrospira platensis biomass with high protein content cultivated in continuous process using urea as nitrogen source. Journal of Applied Microbiology, 112(6), 1086-1094.

5. Messaoudi, O (2020). Isolation and Characterization of Novel Bioactive Molecules from
D'Actinomycètes Isolées du sol Algérien. Ph.D. Thesis, University of Tlemcen, Tlemcen, Algeria,.

6. Baltz, R. H., Demain, A. L., & Davies, J. E. (Eds.). (2010). Manual of industrial microbiology and biotechnology. American Society for Microbiology Press.

7. Messaoudi, O., Wink, J., & Bendahou, M. (2020). Diversity of actinobacteria isolated from date palms rhizosphere and saline environments: Isolation, identification and biological activity evaluation. Microorganisms, 8(12), 1853.

8. Becker, J., & Wittmann, C. (2017). Industrial microorganisms: Corynebacterium glutamicum. Industrial biotechnology: microorganisms, 1, 183-220.

9. Bergmans, HE, van Die, ,I.M., Hoekstra, WP. 1981. Transformation in Escherichia coli: stages in the process. J. Bacteriol. 146:564-570.

10. Bregni, C., Degrossi, J., Garcia, R., Lamas, M. C., Firenstein, R., & D'aquino, M. (2000). Alginate microspheres of Bacillus subtilis.

11. Brindha, J., Ramudu, K. N., & Balamurali, M. M. (2020). An efficient single pot DNA recombination method for protein library generation. International journal of biological macromolecules, 146, 661-667.

12. Cepeda-García, C., Domínguez-Santos, R., García-Rico, R. O., García-Estrada, C., Cajiao, A., Fierro, F., & Martín, J. F. (2014). Direct involvement of the CreA transcription factor in penicillin biosynthesis and expression of the pcbAB gene in Penicillium chrysogenum. Applied microbiology and biotechnology, 98(16), 7113-7124.

13. Chang DE, Jung HC, Rhee JS, Pan JG (1999) Homofermentative production of D()) or L(+)lactate in metabolically engineered Echerichia coli RR1. Appl. Environ. Microbiol. 65: 1384- 1389.

14. Chang, A. Y., Chau, V. W., Landas, J. A., & Pang, Y. (2017). Preparation of calcium competent Escherichia coli and heat-shock transformation. JEMI methods, 1, 22-25.

15. Chawla, J., & Kvarnberg, D. (2014). Hydrosoluble vitamins. Handbook of clinical neurology, 120, 891-914.

16. Choudhary, A. Q., & Pirt, S. J. (1966). The influence of metal-complexing agents on citric acid production by Aspergillus niger. Microbiology, 43(1), 71-81.

17. Christy, P. M., Gopinath, L. R., & Divya, D. (2014). A review on anaerobic decomposition and enhancement of biogas production through enzymes and microorganisms. Renewable and Sustainable Energy Reviews, 34, 167-173.

18. Cirne DG, Lehtomaki A, Bjornsson L, Blackall LL. (2007). Hydrolysis and microbial community analyses in two-stage anaerobic digestion of energy crops. J Appl Microbiol.103: 516- 27.

19. Cvetković, D. D., & Markov, S. L. (2002). Cultivation of tea fungus on malt extract. medium. Acta periodica technologica, (33), 117-124.

20. Darouneh, E., Alavi, A., Vosoughi, M., Arjm, M., Seifkordi, A., & Rajabi, R. (2009). Citric acid production: Surface culture versus submerged culture. African Journal of Microbiology Research, 3(9), 541-545.

21. Bouzidi, A., Azizi, A., Messaoudi, O., Abderrezzak, K., Vidari, G., Hellal, A. N., & Patel, C. N. (2023). Phytochemical analysis, biological activities of methanolic extracts and an isolated flavonoid from Tunisian Limoniastrum monopetalum (L.) Boiss: an in vitro and in silico investigations. Scientific Reports, 13(1), 19144.

22. De La Jara, A., Ruano-Rodriguez, C., Polifrone, M., Assunçao, P., Brito-Casillas, Y., Wägner,
A. M., & Serra-Majem, L. (2018). Impact of dietary Arthrospira (Spirulina) biomass consumption on human health: main health targets and systematic review. Journal of Applied Phycology, 30(4), 2403-2423.

23. Degu, A., Hatew, B., Nunes-Nesi, A., Shlizerman, L., Zur, N., Katz, E & Sadka, A. (2011). Inhibition of aconitase in citrus fruit callus results in a metabolic shift towards amino acid biosynthesis. Planta, 234(3), 501-513.

24. Delaunay, S., Gourdon, P., Lapujade, P., Mailly, E., Oriol, E., Engasser, J. M., ... & Goergen, J.
L. (1999). An improved temperature-triggered process for glutamate production with Corynebacterium glutamicum. Enzyme and microbial technology, 25(8-9), 762-768.

25. Demain, A. L. (1981). Industrial microbiology. Science, 214(4524), 987-995.

26. Demain, A. L., & Adrio, J. L. (2008). Contributions of microorganisms to industrial biology. Molecular biotechnology, 38(1), 41.

27. Devi, P., D'Souza, L., Kamat, T., Rodrigues, C., & Naik, C. G. (2009). Batch culture fermentation of Penicillium chrysogenum and a report on the isolation, purification, identification and antibiotic activity of citrinin.

28. Diethard, M., Gasser, B., Egermeier, M., Marx, H., & Sauer, M. (2017). Industrial Microorganisms: Saccharomyces cerevisiae and other Yeasts. Industrial Biotechnology: Microorganisms, 2, 673-686.

29. Dumitriu, S. (Ed.). (2004). Polysaccharides: structural diversity and functional versatility. CRC press.

30. Dziegielewska-Gesiak, S., Fatyga, E., Kasiarz, G., Wilczynski, T., Muc-Wierzgon, M., & Kokot, T. (2019). Vitamins B12 and D deficiencies and macro-and microelement disturbances among diabetic elderly patients. Journal of biological regulators and homeostatic agents, 33(2), 477- 483.

31. El Khoury, J. (2019). Investigation of B-lactam resistance in Streptococcus pneumoniae, Escherichia coli, Klebsiella pneumoniae and Pseudomonas aeruginosa using omics approaches.

32. Elefsiniotis P, Oldham WK. 2004. Anaerobic acidogenesis of primary sludge: the role of solids retention time. Biotechnol Bioeng;44: 7-13.

33. El-Mansi, E. M. T., Nielsen, J., Mousdale, D., & Carlson, R. P. (Eds.). (2018). Fermentation microbiology and biotechnology. CRC press.

34. El-Nawwi, S. A., & Abd El-Kader, A. (1996). Production of single-cell protein and cellulase from sugarcane bagasse: effect of culture factors. Biomass and bioenergy, 11(4), 361-364.

35. Enzmann, F. Mayer, M. Rother, D (2018). Holtmann Methanogens: biochemical background and biotechnological applications AMB Express, 8, 1.

36. Erickson. L.E (2011). Bioreactors for Commodity Products. Editor(s): Murray Moo-Young, Comprehensive Biotechnology (Second Edition), Academic Press,. 653-658.

37. Erickson. L.E (2019). Bioreactors, Editor(s): Thomas M. Schmidt, Encyclopedia of Microbiology (Fourth Edition), Academic Press. 536-541.

38. Ertan, H. (1992). Some properties of glutamate dehydrogenase, glutamine synthetase and glutamate synthase from Corynebacterium callunae. Archives of microbiology, 158(1), 35-41.

39. Evgenios K, Christos C (2020). Model-based dynamic optimization of the fermentative

production of polyhydroxyalkanoates (PHAs) in fed-batch and sequence of continuously operating bioreactors. Biochemical Engineering Journal.162: 107702.

40. Fabregas, J., & Herrero, C. (1985). Marine microalgae as a potential source of single cell protein (SCP). Applied microbiology and biotechnology, 23(2), 110-113.

41. Messaoudi, O., Sudarman, E., Patel, C., Bendahou, M., & Wink, J. (2022). Metabolic profile, biotransformation, docking studies and molecular dynamics simulations of bioactive compounds secreted by CG3 strain. *Antibiotics*, *11*(5), 657.

42. Ferreira, A. F., Ribeiro, A. M., Kulaç, S., & Rodrigues, A. E. (2015). Methane purification by adsorptive processes on MIL-53 (Al). Chemical Engineering Science, 124, 79-95.

43. Francioso, O., Rodriguez-Estrada, M. T., Montecchio, D., Salomoni, C., Caputo, A., & Palenzona, D. (2010). Chemical characterization of municipal wastewater sludges produced by two- phase anaerobic digestion for biogas production. Journal of hazardous materials, 175(1-3), 740-746.

44. Franklin, M. J., Nivens, D. E., Weadge, J. T., & Howell, P. L. (2011). Biosynthesis of the Pseudomonas aeruginosa extracellular polysaccharides, alginate, Pel, and Psl. Frontiers in microbiology, 2, 167.

45. Garcıa-Ochoa, F., Santos, V. E., Casas, J. A., & Gómez, E. (2000). Xanthan gum: production, recovery, and properties. Biotechnology advances, 18(7), 549-579.

46. Gervasi, T., Pellizzeri, V., Calabrese, G., Di Bella, G., Cicero, N., & Dugo, G. (2018). Production of single cell protein (SCP) from food and agricultural waste by using Saccharomyces cerevisiae. Natural product research, 32(6), 648-653.

47. Ghanem (1992). Single cell protein production from beet pulp by mixed culture.

48. Gumus, T., Demirci, A.S., Mirik, M., Arici, M., Aysan, Y., 2010. Xanthan gum production of Xanthomonas spp. isolated from different plants. Food Science Biotechnology 19 (1), 201-206.

49. Gutmann, M., Hoischen, C., & Krämer, R. (1992). Carrier-mediated glutamate secretion by Corynebacterium glutamicum under biotin limitation. Biochimica et Biophysica Acta (BBA)- Biomembranes, 1112(1), 115-123.

50. Hamidi, M., Kennedy, J. F., Khodaiyan, F., Mousavi, Z., & Hosseini, S. S. (2019). Production optimization, characterization and gene expression of pullulan from a new strain of Aureobasidium pullulans. International journal of biological macromolecules, 138, 725-735.

51. Harwood, C. R., Park, S. H., & Sauer, M. (2018). Editorial for the thematic issue on "Industrial Microbiology".

52. Hazra, A. B., Han, A. W., Mehta, A. P., Mok, K. C., Osadchiy, V., Begley, T. P., &

Taga, M. E. (2015). Anaerobic biosynthesis of the lower ligand of vitamin B12. Proceedings of the National Academy of Sciences, 112(34), 10792-10797.

53. He, F., Yang, Y., Yang, G., & Yu, L. (2010). Studies on antibacterial activity and antibacterial mechanism of a novel polysaccharide from Streptomyces virginia H03. Food Control, 21(9), 1257- 1262.

54. Hermann, T. (2003). Industrial production of amino acids by coryneform bacteria. Journal of biotechnology, 104(1-3), 155-172.

55. Hirasawa, T., & Wachi, M. (2016). Glutamate fermentation-2: mechanism of L-glutamate overproduction in Corynebacterium glutamicum. Amino Acid Fermentation, 57-72.

56. Hoischen C, Kra¨mer R (1990) Membrane alteration is necessary but not sufficient for effective glutamate secretion in Corynebacterium glutamicum. J Bacteriol 172(6):3409-3416.

57. Huang, C., Luo, M. T., Chen, X. F., Xiong, L., Li, X. M., & Chen, X. D. (2017). Recent advances and industrial viewpoint for biological treatment of wastewaters by oleaginous microorganisms. Bioresource technology, 232, 398-407.

58. Huang, X., Liu, X., Chen, F., Wang, Y., Li, X., Wang, D., ... & Yang, Q. (2020). Clarithromycin affects methane production from anaerobic digestion of waste activated sludge. Journal of Cleaner Production, 255, 120321.

59. Humphrey, A. E., & Lee, S. E. (1992). Industrial fermentation: principles, processes, and products. In Riegel's handbook of industrial chemistry (pp. 916-986). Springer, Dordrecht.

60. Ikeda, M. (2003). Amino acid production processes. Microbial production of l-amino acids, 1- 35.

61. Ikeda, M., & Takeno, S. (2013). Amino acid production by Corynebacterium glutamicum. In Corynebacterium glutamicum (pp. 107-147). Springer, Berlin, Heidelberg.

62. Ikram-Ul, H., Ali, S., Qadeer, M. A., & Iqbal, J. (2004). Citric acid production by selected mutants of Aspergillus niger from cane molasses. Bioresource Technology, 93(2), 125-130.

63. Jang, C, Magnuson, T. 2013. A Novel Selection Marker for Efficient DNA Cloning and Recombineering in E. coli. Plos One. 8:e57075.

64. Jiang, L. (2013). Effect of nitrogen source on curdlan production by Alcaligenes faecalis ATCC 31749. International journal of biological macromolecules, 52, 218-220.

65. Johansen, E. (2017). Future access and improvement of industrial lactic acid bacteria cultures. Microbial cell factories, 16(1), 1-5.

66. Kalai, S., Bensoussan, M., Dantigny, P. (2014). Lag time for germination of

Penicillium chrysogenum conidia is induced by temperature shifts. Food Microbiology, 42, 149-153.

67. Kampen, W. H. (2014). Nutritional requirements in fermentation processes. In Fermentation and biochemical engineering handbook (pp. 37-57). William Andrew Publishing.

68. Keller, N. P. (2019). Fungal secondary metabolism: regulation, function and drug discovery. Nature Reviews Microbiology, 17(3), 167-180.

69. Khan KH, Shaukat SS (1990). Citric acid production with mixed strains of Aspergillus niger in submerged culture. Acta Microbiol. Hung 37: 9-13.

70. Kılıç, M., Bayraktar, E., Ateş, S., & Mehmetoglu, Ü. (2002). Investigation of extractive citric
acid fermentation using response-surface methodology. Process Biochemistry, 37(7), 759-767.

71. Kim, J., Fukuda, H., Hirasawa, T., Nagahisa, K., Nagai, K., Wachi, M., & Shimizu, H. (2010). Requirement of de novo synthesis of the OdhI protein in penicillin-induced glutamate production by Corynebacterium glutamicum. Applied microbiology and biotechnology, 86(3), 911-920.

72. Kim, J., Hirasawa, T., Sato, Y., Nagahisa, K., Furusawa, C., & Shimizu, H. (2009). Effect of odhA overexpression and odhA antisense RNA expression on Tween-40-triggered glutamate production by Corynebacterium glutamicum. Applied microbiology and biotechnology, 81(6), 1097-1106.

73. Koval, O., & Oliinichuk, S. (2019). Influence of melasses non-sugars on efficiency of fermentation of mash from sacchariferous raw material.

74. Kräutler, B. (2005). Vitamin B12: chemistry and biochemistry. Biochemical Society Transactions, 33(4), 806-810.

75. Kriger, O. V., & Noskova, S. Y. (2018). Properties of lactic acid microorganisms: long-term preservation methods. Food Processing: Techniques and Technology, 51(4), 30-38.

76. Krishnakumar, V., Seshadri, S., & Muthunatasen, S. (2007). Analysis of vibrational spectra of 5, 6-dimethyl benzimidazole based on density functional theory calculations. Spectrochimica Acta Part A: Molecular and Biomolecular Spectroscopy, 68(3), 811-816.

77. Kubicek, C. P., & Röhr, M. (1985). Aconitase and citric acid fermentation by Aspergillus niger. Applied and environmental microbiology, 50(5), 1336-1338.

78. Kushkevych, I., Kobzová, E., Vítězová, M., Vítěz, T., Dordević, D., & Bartoš, M. (2019). Acetogenic microorganisms in operating biogas plants depending on substrate combinations. Biologia, 74(9), 1229-1236.

79. Laluce, C., Bertolini, M. C., Ernandes, J. R., Martini, A. V., & Martini, A. (1988). New

amylolytic yeast strains for starch and dextrin fermentation. Applied and environmental microbiology, 54(10), 2447-2451.

80. Lee, K. S., Boccazzi, P., Sinskey, A. J., & Ram, R. J. (2011). Microfluidic chemostat and turbidostat with flow rate, oxygen, and temperature control for dynamic continuous culture. Lab on a Chip, 11(10), 1730-1739.

81. Leela, J. K., & Sharma, G. (2000). Studies on xanthan production from Xanthomonas campestris. Bioprocess Engineering, 23(6), 687-689.

82. Lehtomäki, A. (2006). Biogas production from energy crops and crop residues (No. 163). University of Jyväskylä.

83. Lyu, Z., Shao, N., Akinyemi, T., & Whitman, W. B. (2018). Methanogenesis. Current Biology, 28(13), R727-R732.

84. Martin, M. R., Fornero, J. J., Stark, R., Mets, L., & Angenent, L. T. (2013). A single-culture bioprocess of Methanothermobacter thermautotrophicus to upgrade digester biogas by CO2-to-CH4 conversion with H2. Archaea, 2013.

85. Becheur, M., Lounas, A., Messaoudi, O., Oumouna-Benachour, K., & Oumouna, M. (2024). molecular characterization of infectious bursal disease virus isolated from broiler and pullet flocks in algeria. journal of animal and plant sciences-japs, 34(1), 73-89.

86. Max, B., Salgado, J. M., Rodríguez, N., Cortés, S., Converti, A., & Domínguez, J. M. (2010). Biotechnological production of citric acid. Brazilian journal of Microbiology, 41(4), 862-875.

87. Menzel, C., Olsson, E., Plivelic, T. S., Andersson, R., Johansson, C., Kuktaite, R. & Koch, K. (2013). Molecular structure of citric acid cross-linked starch films. Carbohydrate polymers, 96(1), 270-276.

88. Min, B. E., Hwang, H. G., Lim, H. G., & Jung, G. Y. (2017). Optimization of industrial microorganisms: recent advances in synthetic dynamic regulators. Journal of Industrial Microbiology and Biotechnology, 44(1), 89-98.

89. Naessens, M., Cerdobbel, A. N., Soetaert, W., & Vandamme, E. J. (2005). Leuconostoc dextransucrase and dextran: production, properties and applications. Journal of Chemical Technology & Biotechnology: International Research in Process, Environmental & Clean Technology, 80(8), 845-860.

90. Nampoothiri, K. M., Singhania, R. R., Sabarinath, C., & Pandey, A. (2003). Fermentative production of gellan using Sphingomonas paucimobilis. Process Biochemistry, 38(11), 1513-1519.

91. Nangul, A., & Bhatia, R. (2021). Microorganisms: a marvelous source of single cell proteins. Journal of Microbiology, Biotechnology and Food Sciences, 2021, 15-18.

92. Nasseri, A. T., Rasoul-Amini, S., Morowvat, M. H., & Ghasemi, Y. (2011). Single cell

protein: production and process. American Journal of food technology, 6(2), 103-116.

93. Nunez, C., Pena, C., Kloeckner, W., Hernández-Eligio, A., Bogachev, A.V., Moreno, S., Guzmán, J., Büchs, J., Espín, G., 2013. Alginate synthesis in Azotobacter vinelandii is increased by reducing the intracellular production of ubiquinone. Appl. Microbiol. Biotechnol. 97, 2503-2512.

94. Messaoudi, O. Contribution à la Caractérisation des Souches D'Actinomycètes Productrice des Métabolites Antibactériennes Isolée de la Sebkha de Kendasa (Bechar). Master's Thesis, University of Tlemcen, Tlemcen, Algeria, 2013.

95. Messaoudi, O., Bendahou, M., Benamar, I., & Abdelwouhid, D. E. (2015). Identification and preliminary characterization of non-polyene antibiotics secreted by new strain of actinomycete isolated from sebkha of Kenadsa, Algeria. *Asian pacific journal of Tropical biomedicine*, *5*(6), 438- 445.

96. Oosterhuis, N.M.G. Hudson, T. D'Avino, A. Zijlstra, G.M. Amanullah, A (2011). Disposable Bioreactors, Editor(s): Murray Moo-Young, Comprehensive Biotechnology (Second Edition), Academic. 249-261.

97. Page, M. G. (2012). Beta-lactam antibiotics. In Antibiotic Discovery and Development (pp. 79- 117). Springer, Boston, MA.

98. Palaniraj, A., & Jayaraman, V. (2011). Production, recovery and applications of xanthan gum by Xanthomonas campestris. Journal of Food Engineering, 106(1), 1-12.

99. Messaoudi, O., Gouzi, H., El-Hoshoudy, A. N., Benaceur, F., Patel, C., Goswami, D & Bendahou, M. (2021). Berries anthocyanins as potential SARS-CoV-2 inhibitors targeting the viral attachment and replication; molecular docking simulation. *Egyptian Journal of Petroleum*, *30*(1), 33-43.

100. Papagianni, M. (2007). Advances in citric acid fermentation by Aspergillus niger: biochemical aspects, membrane transport and modeling. Biotechnology advances, 25(3), 244-263.

101. Papagianni, M., Wayman, F., & Mattey, M. (2005). Fate and role of ammonium ions during fermentation of citric acid by Aspergillus niger. Applied and environmental microbiology, 71(11), 7178-7186.

102. Parmar, A., Kumar, H., Marwaha, S. S., & Kennedy, J. F. (2000). Advances in enzymatic transformation of penicillins to 6-aminopenicillanic acid (6-APA). Biotechnology advances, 18(4), 289-301.

103. Paul, P. E. V., Sangeetha, V., & Deepika, R. G. (2019). Emerging trends in the industrial production of chemical products by microorganisms. In Recent developments in applied microbiology and biochemistry (pp. 107-125). Academic Press.

104. Pelton, R. (2002). A review of antifoam mechanisms in fermentation. Journal of

industrial microbiology and biotechnology, 29(4), 149-154.

105. Pescuma, M., de Valdez, G. F., & Mozzi, F. (2015). Whey-derived valuable products obtained by microbial fermentation. Applied microbiology and biotechnology, 99(15), 6183-6196.

106. Piao, Y., Yamashita, M., Kawaraichi, N., Asegawa, R., Ono, H., & Murooka, Y. (2004). Production of vitamin B12 in genetically engineered Propionibacterium freudenreichii. Journal of bioscience and bioengineering, 98(3), 167-173.

107. Piwowarek, K., Lipińska, E., Hać-Szymańczuk, E., Kieliszek, M., & Ścibisz, I. (2018). Propionibacterium spp.-source of propionic acid, vitamin B12, and other metabolites important for the industry. Applied microbiology and biotechnology, 102(2), 515-538.

108. Porter, N. T., & Martens, E. C. (2017). The critical roles of polysaccharides in gut microbial ecology and physiology. Annual review of microbiology, 71, 349-369.

109. Purama, R. K., Goswami, P., Khan, A. T., & Goyal, A. (2009). Structural analysis and properties of dextran produced by Leuconostoc mesenteroides NRRL B-640. Carbohydrate Polymers, 76(1), 30-35.

110. Ramakrishnan, C. V., Steel, R., & Lentz, C. P. (1955). Mechanism of citric acid formation and accumulation in Aspergillus niger. Archives of biochemistry and biophysics, 55(1), 270-273.

111. Ravindra, P. (2000). Value-added food:: Single cell protein. Biotechnology advances, 18(6), 459-479.

112. Reihani, S. F. S., & Khosravi-Darani, K. (2019). Influencing factors on single-cell protein production by submerged fermentation: A review. Electronic Journal of Biotechnology, 37, 34-40.

113. Ritala, A., Häkkinen, S. T., Toivari, M., & Wiebe, M. G. (2017). Single cell protein-state-of- the-art, industrial landscape and patents 2001-2016. Frontiers in microbiology, 8, 2009.

114. Robyt, J. F., Yoon, S. H., & Mukerjea, R. (2008). Dextransucrase and the mechanism for dextran biosynthesis. Carbohydrate research, 343(18), 3039-3048.

115. Messaoudi, O., Steinmann, E., Praditya, D., Bendahou, M., & Wink, J. (2022). Taxonomic characterization, antiviral activity and induction of three new Kenalactams in Nocardiopsis sp. CG3. Current Microbiology, 79(9), 284.

116. Rosalam, S., & England, R. (2006). Review of xanthan gum production from unmodified starches by Xanthomonas comprestris sp. Enzyme and Microbial Technology, 39(2), 197-207.

117. Sagagi, B., Garba, B., & Usman, N. (2009). Studies on biogas production from fruits and vegetable waste. Bayero Journal of Pure and Applied Sciences, 2(1), 115-118.

118. Salam, N., Jiao, J. Y., Zhang, X. T., & Li, W. J. (2020). Update on the classification of higher ranks in the phylum Actinobacteria. International journal of systematic and evolutionary microbiology, 70(2), 1331-1355.

119. Salehizadeh, H., Yan, N., & Farnood, R. (2018). Recent advances in polysaccharide bio-based flocculants. Biotechnology advances, 36(1), 92-119.

120. Sarkar, D., & Modak, J. M. (2003). Optimisation of fed-batch bioreactors using genetic algorithms. Chemical Engineering Science, 58(11), 2283-2296.

121. Schink, B (1997). Energetics of syntrophic cooperation in methanogenic degradation. Microb Mol Biol Rev 61:262-280.

122. Schultz, C., Niebisch, A., Gebel, L., & Bott, M. (2007). Glutamate production by Corynebacterium glutamicum: dependence on the oxoglutarate dehydrogenase inhibitor protein OdhI and protein kinase PknG. Applied microbiology and biotechnology, 76(3), 691-700.

123. Shao, W., Ma, K., Le, Y., Wang, H., & Sha, C. (2017). Development and use of a novel random mutagenesis method: in situ error-prone PCR (is-epPCR). In In Vitro Mutagenesis (pp. 497-506). Humana Press, New York, NY.

124. Messaoudi, O., Sudarman, E., Bendahou, M., Jansen, R., Stadler, M., & Wink, J. (2019). Kenalactams A-E, polyene macrolactams isolated from Nocardiopsis CG3. Journal of Natural Products, 82(5), 1081-1088.

125. Show, P. L., Oladele, K. O., Siew, Q. Y., Aziz Zakry, F. A., Lan, J. C. W., & Ling, T. C. (2015). Overview of citric acid production from Aspergillus niger. Frontiers in life science, 8(3), 271-283.

126. Silva, M.F., Fornari, R.C.G., Mazutti, M.A., Oliveira, D., Padilha, F.F., Cichoski, A.J., Cansian, R.L., Luccio, M.D., Treichel, H., 2009. Production and characterization of xanthan gum by Xanthomonas campestris using cheese whey as sole carbon source. Journal of Food Engineering 90, 119-123.

127. Singh, R. S., Saini, G. K., & Kennedy, J. F. (2008). Pullulan: microbial sources, production and applications. Carbohydrate polymers, 73(4), 515-531.

128. Sonenshein, A. L. (2001). The Krebs citric acid cycle. Bacillus subtilis and its Closest Relatives: from Genes to Cells, 151-162.

129. Steiger, M. G., Rassinger, A., Mattanovich, D., & Sauer, M. (2019). Engineering of the citrate exporter protein enables high citric acid production in Aspergillus niger. Metabolic engineering, 52, 224-231.

130. Suarez, C., & Gudiol, F. (2009). Beta-lactam antibiotics. Enfermedades infecciosas y microbiologia clinica, 27(2), 116-129.

131. Sworn, G. (2021). Xanthan gum. In Handbook of hydrocolloids (pp. 833-853).

Woodhead Publishing.

132. Szczodrak, J. (1981). Biosynthesis of citric acid in relation to the activity of selected enzymes of the Krebs cycle in Aspergillus niger mycelium. European journal of applied microbiology and biotechnology, 13(2), 107-112.

133. Tatsumi, N., & Inui, M. (Eds.). (2012). Corynebacterium glutamicum: biology and biotechnology (Vol. 23). Springer Science & Business Media.

134. Tian, G., Zhang, W., Dong, M., Yang, B., Zhu, R., Yin, F., ... & Cui, X. (2017). Metabolic pathway analysis based on high-throughput sequencing in a batch biogas production process. Energy, 139, 571-579.

135. Tiwari, S., Jamal, S. B., de Carvalho, P. V. S. D., Hassan, S. S., Silva, A., & Azevedo, V. Industrial Microbiology & Biotechnology.

136. Vandenberghe, L. P., Soccol, C. R., Pandey, A., & Lebeault, J. M. (1999). Microbial production of citric acid. Brazilian Archives of Biology and Technology, 42(3), 263-276.

137. Waites, M. J., Morgan, N. L., Rockey, J. S., & Higton, G. (2009). Industrial microbiology: an introduction. John Wiley & Sons.

138. Walker, G. M. (2014). Fermentation (Industrial): media for industrial fermentations. In Encyclopedia of food microbiology (pp. 769-777). Academic Press.

139. Wilson, D. B., Sahm, H., Stahmann, K. P., & Koffas, M. (Eds.). (2019). Industrial Microbiology. John Wiley & Sons.

140. Xu, H., Jiang, M., Li, H., Lu, D., & Ouyang, P. (2005). Efficient production of poly (y-glutamic acid) by newly isolated Bacillus subtilis NX-2. Process Biochemistry, 40(2), 519-523.

141. Yalcin, S. K., Bozdemir, M. T., & Ozbas, Z. Y. (2010). Citric acid production by yeasts: fermentation conditions, process optimization and strain improvement. Current research, technology and education topics in applied microbiology and microbial biotechnology, 9, 1374- 1382.

142. Yamauchi, R., Maguin, E., Horiuchi, H., Hosokawa, M., & Sasaki, Y. (2019). The critical role of urease in yogurt fermentation with various combinations of Streptococcus thermophilus and Lactobacillus delbrueckii ssp. bulgaricus. Journal of dairy science, 102(2), 1033-1043.

143. Yang, H., Swartz, A. M., Park, H. J., Srivastava, P., Ellis-Guardiola, K., Upp, D. M & Lewis,
J. C. (2018). Evolving artificial metalloenzymes via random mutagenesis. Nature chemistry, 10(3), 318.

144. Yang, M. J., Lee, H. W., & Kim, H. (2017). Enhancement of thermostability of Bacillus subtilis endoglucanase by error-prone PCR and DNA shuffling. Applied Biological

Chemistry, 60(1), 73-78.

145. Ye, B., Li, Y., Tao, Q., Yao, X., Cheng, M., & Yan, X. (2020). Random mutagenesis by insertion of error-prone PCR products to the chromosome of Bacillus subtilis. Frontiers in Microbiology, 11.

145. Yongsmith, B., Sonomoto, K., Tanaka, A., & Fukui, S. (1982). Production of vitamin B 12 by immobilized cells of a propionic acid bacterium. European journal of applied microbiology and biotechnology, 16(2), 70-74.

146. Zähner, H., Anke, H., & Anke, T. (2020). Evolution and secondary pathways. In Secondary metabolism and differentiation in fungi (pp. 153-171). CRC Press.

147. Zhang, B., Jiang, Y., Li, Z., Wang, F., & Wu, X. Y. (2020). Recent Progress on Chemical Production From Non-food Renewable Feedstocks Using Corynebacterium glutamicum. Frontiers in Bioengineering and Biotechnology, 8.

148. Zhang, W., & Nielsen, D. (2014). Synthetic biology applications in industrial microbiology. Frontiers in microbiology, 5, 451.

149. Zhou, S., Du, G., Kang, Z., Li, J., Chen, J., Li, H., & Zhou, J. (2017). The application of powerful promoters to enhance gene expression in industrial microorganisms. World Journal of Microbiology and Biotechnology, 33(2), 23.

150. Zhou, S., Shanmugam, K. T., Yomano, L. P., Grabar, T. B., & Ingram, L. O. (2006). Fermentation of 12%(w/v) glucose to 1.2 M lactate by Escherichia coli strain SZ194 using mineral salts medium. Biotechnology letters, 28(9), 663-670.

MIX
Papier aus verantwortungsvollen Quellen
Paper from responsible sources
FSC® C105338

Printed by Books on Demand GmbH, Norderstedt / Germany